The Map in the Machine

The Map in the Machine

Charting the Spatial Architecture of Digital Capitalism

Luis F. Alvarez Leon

UNIVERSITY OF CALIFORNIA PRESS

University of California Press
Oakland, California

Library of Congress Cataloging-in-Publication Data

Names: Alvarez León, Luis F., author.
Title: The map in the machine : charting the spatial architecture
    of digital capitalism / Luis F. Alvarez Leon.
Description: Oakland, California : University of California
    Press, [2024] | Includes bibliographical references and index.
Identifiers: LCCN 2023048026 (print) | LCCN 2023048027
    (ebook) | ISBN 9780520389304 (cloth) | ISBN 9780520389328
    (paperback) | ISBN 9780520389335 (epub)
Subjects: LCSH: Mobile geographic information systems—
    Economic aspects.
Classification: LCC G109.4 .A58 2024 (print) | LCC G109.4
    (ebook) | DDC 910.285—dc23/eng/20231122
LC record available at https://lccn.loc.gov/2023048026
LC ebook record available at https://lccn.loc.gov/2023048027

33 32 31 30 29 28 27 26 25 24
10 9 8 7 6 5 4 3 2 1

For Tikal, without whose interludes from barking
this book would not have been written.

Contents

List of Illustrations and Table

Acknowledgments

In more ways than one this book is like the proverbial rug that ties the room together. When editor extraordinaire Michelle Lipinski, from UC Press, contacted me with the idea to write a monograph, I had to think about whether I had a story to tell in this format. After considering various preliminary ideas, I realized that the story I was looking for was not in a single, standalone narrative, but rather in the interstices between various research projects. This book then became the place to articulate the question that motivates my entire scholarship: to understand the spatial architecture of digital capitalism. Fueled by that insight, I forged ahead to write a book that both synthesizes past and present lines of research and offers what I hope is a cohesive (and hopefully coherent) perspective on how geographic data, media, and technologies play a fundamental role in the workings of digital capitalism.

Since this is a volume that threads narratives that began several years ago in graduate school and have followed me through a postdoc, a visiting position, and now through the fifth year of a tenure-track appointment, gratitude for every step of that journey permeates each word and each page. While I may have typed the words in the following pages, in fundamental ways they are the product of the multiple communities that have shaped me, to whom I owe an unpayable debt, and to whom I hope I can contribute in some way. Yet, while this book is the work of an extended and ever-growing village (that is starting to look like a city), the so-called death of the author cannot absolve me from ultimate responsibility for what is contained within these pages.

First of all, my sincere thanks to Michelle Lipinski, senior editor, and the rest of the wonderful editorial team at University of California Press: Jyoti Arvey, LeKeisha Hughes, Enrique Ochoa-Kaup, and the rest of the staff. To the anonymous referees who reviewed my proposal and book manuscript, I am very grateful for the care and attention with which you engaged with my work. Thanks also to Jon Dertien, Gary Hamel, and the team at BookComp for their outstanding work in the production of this book.

As I will inevitably, and unconscionably, miss some names in what follows, I must say to those have been with me in any capacity along this journey: you know who you are.

While an academic career is an exercise in finding one's own identity through the asking (and hopefully answering) of questions, it bears the permanent imprint of those who advise us at key stages in our formation. For that indelible intellectual imprint, I thank my doctoral advisors, Allen J. Scott and Eric S. Sheppard. I also thank my PhD cohort and friends from the UCLA Geography Department, who welcomed me into their community and encouraged my first steps into the world of geography.

Friends at the Sol Price Center for Social Innovation at USC, Gary Painter, Jung Hyun Choi, Soledad Gregorio, Elly Schoen, Sean Angst, Caroline Bhalla, Megan Goulding, and the rest of the team who worked on the Neighborhood Data for Social Change platform. Special thanks to friend and coauthor Jovanna Rosen, who played a vital role in developing many key ideas in this book and offered truly excellent feedback on the manuscript, for which I am immensely grateful.

Colleagues at Clark University Graduate School of Geography and in particular coauthor Yuko Aoyama, from whom I've learned a lot through our joint projects, and whose work has influenced the development of my own intellectual trajectory.

Friends, colleagues, and students at the Dartmouth College Department of Geography and beyond. Thank you to Abigail Neely, Chris Sneddon, Coleen Fox, Erin Collins, Frank Magilligan, Jonathan Chipman, Jonathan Winter, Justin Mankin, Mona Domosh, Richard Wright, Tish Lopez, Susanne Freidberg, Xun Shi, Jane Henderson, Aletha Spang, Alex Bramsen, Maron Greenleaf, and Kelly Palmer (and Birdie). The guidance and support of faculty mentors has been invaluable to my growth as a teacher and scholar, so additionally I would like to thank Emily Walton for her advice and encouragement. Special thanks to the postdocs, researchers, and students who have worked at the Critical Geospatial Research Lab for the past three years: Meghan Kelly, Park Muhonda, Janice Kai Chen, Catherine Granville, Ishika Jha, Madeleine Morris, Rocío Barrionuevo Quispe, Rothschild Toussaint, Sheen Kim, and Gavan Fink.

At the AAG Digital Geographies Specialty Group I would like to acknowledge the collective work of Eric Robsky Huntley, Will Payne, Emma Fraser, Clancy Wilmott, Ryan Burns, Ian Spangler, Shiloh Deitz, and Will Payne in particular.

Thank you also to the members of the AAG Economic Geography Specialty Group for their support of my work over the years. In particular, thank you to Abigail Cooke, Emily Rosenman, Peter Kedron, and Jennifer Clark.

The Matrix Group in Digital Transformations in Land, Property, and Housing has been a source of intellectual inspiration, friendship, and camaraderie. Thank you to Desiree Fields, Hilary Faxon, Jovanna Rosen (again), Julien Migozzi, Courtney Wittekind, Elizabeth Resor, Kendra Kintzi, and (Cartus) Bo-Xiang You.

Friends and colleagues: Matt Zook, Jeremy Crampton, Sarah Elwood, Jamie Peck, Matt Wilson, Sterling Quinn, Colin Gleason, Mia Bennett, Jenny Goldstein, Eric Nost, Johannes Glückler, Garrett Dash Nelson, Ben Gerlofs, Emma Colven, Sam Nowak, Sophie Webber, Sierra Burkhart, Michael Shin, Cameran Ashraf, Sara Hughes, Clare Beer, Sean Kennedy, Dylan Connor, Tyler Harlan, Dimitar Anguelov, Dillon Mahmoudi, Craig Dalton, Jim Thatcher, Renee Tapp, Alan Wiig, Alex Tarr, folks at the Oxford Internet Institute Summer Doctoral Program 2015, and the Summer Institute in Economic Geography 2014, and many, many more.

Thank you to Fulbright-García Robles, UC MEXUS-CONACYT, USC-CONACYT, and Dartmouth Burke Award for funding significant portions of the research that went into writing this book.

My deepest gratitude to the León Manríquez Family, and in particular Yolanda León Manríquez for all the love, support, and inspiration at every step of the way, Fernanda Alvarez León (always the duck connection), José Luis León Manríquez (mentor and colleague), and Carlos Quiroz (for all the support, estilo Tamaulipas). To Luis Alvarez Colín, I wish you could have read this in real time, but I'll wait for the book review on the other side.

To the van Loggerenberg family, thank you for staying close, even across continents and oceans.

To the Chavos, started in Cerro del Hombre, and now we're here (and there).

Lastly, and most importantly, to Aleksandra, my partner in crime in this grand caper we call life. Siempre.

CHAPTER 1

Introduction

DIGITAL TRANSFORMATIONS

Over the first two decades of the twenty-first century the daily experience of buying and selling products and services has changed drastically for billions of people all over the world. During this period, the digital economy has expanded from primarily software, video games, and other computer applications to encompass vast and varied markets that deliver digitized media via the internet, such as music or video streaming, as well as a broad array of goods and services accessed through digital platforms, often through mobile devices. It is true that before this expansion, in the late 1990s and early 2000s, users were already able to download many types of multimedia files from websites or via free peer-to-peer file sharing networks, such as Napster and LimeWire. However, these services were limited by factors such as narrow bandwidth and inadequate online distribution systems, as well as the shadow of piracy and the threat of legal penalties, which loomed large in peer-to-peer digital exchanges. However, since that time, technical, legal, and societal developments have created conditions for the emergence of "legitimate marketplaces" where providers of digital goods have become not only household names but powerful players in the global economy.

Thus, while file sharing networks and other ways of accessing free multimedia content online have not fully disappeared, they have largely given up ground to corporations like Netflix; Amazon (and its video and music streaming services); Apple (along with its app and music stores and Apple TV+); and a growing number of streaming services that have emerged in the so-called "streaming wars" of the early 2020s: Disney+, NBC's Peacock, HBO+, CBS All Access, and Discovery+, to name a few. Likewise, the publishing industry has been upended not only through the sale of print books online (most notably via Amazon), but also through the widespread popularity of electronic books and e-readers, such as Amazon's Kindle, Barnes and Noble's nook, and Rakuten's Kobo. Parallel to the development of marketplaces for established multimedia products, new ways of accessing products and services through digital means have emerged. Central among these are digital platforms such as Uber, Lyft, Airbnb, Grubhub, Instacart, and countless others, which have leveraged digital tools like smartphones equipped with GPS tracking to provide new kinds of intermediary services that have come to reshape important industries beyond media, such as transportation, real estate, finance, logistics, retail, advertising, and a massive (state and nonstate) surveillance apparatus that bleeds into all of the above.

As the digital economy continues to grow along with the technologies that enable it, a profound transformation related to the digitization of global capitalism is underway. This transformation affects not only the very nature of goods and services that we consume, but also the markets where they circulate, how those markets are constituted, and how people interact with them. This book argues that the digitization of global capitalism has also resulted in profound impacts on the very geographic organization of the economy. In other words, digitization is not only changing how we shop, live, work, play, and communicate, but in the process, it is also reconstituting the very spaces and places where we engage in all these activities, the interconnections between them, and consequently the way we conceive of, organize, and navigate our societies and economies. Initially these changes became most immediately visible in media markets as these were first reoriented toward the trade in digital goods and then largely subsumed by digital networks and platforms. For instance, in the span of the first decade of the twenty-first century, multimedia consumption shifted from CDs and DVDs to individual song or movie purchases via online stores such as Apple's iTunes store, to online subscription streaming services such as Netflix and Spotify. One visible effect of this has been the decimation of brick-and-mortar video stores and music shops, with giants of an earlier era like Blockbuster and Tower Records facing near-total extinction.[1] A less visible effect of the same shift in the media industries has been the reshuffling of power relations between different actors in the production and distribution process (e.g., record labels, artists, retailers, and technology companies). For instance, in May 2023, negotiations for a new contract between the Alliance of Motion Picture and Television Producers (AMPTP) and the Writers Guild of America (WGA) broke down. As a result, the WGA went on strike over changes stemming from the widespread digitization of the film and TV industry and its shift toward streaming, such as lack of residual payment, shorter TV seasons, and the threat of artificial intelligence (AI) as a screenwriting replacement.[2] A few months later, the writers were joined by the Screen Actors Guild (SAG), whose contract negotiations broke down over a similar set of issues involving streaming platform residuals and the use of artificial intelligence to replace labor. At the time of writing, film and television production in the United States has been largely stopped due to the historic double strike paralyzing Hollywood, something not seen since 1960, when Ronald Reagan was the leader of SAG.[3] However, while these changes have profoundly affected media industries, the digital transformation of capitalism goes well beyond the replacement of storefronts with digital portals or shifting power relations between guilds. Rather, it entails a fundamental reshaping of markets for goods and services across sectors of the economy according to a new spatial and technological order—what I call in this book the spatial architecture of digital capitalism.

Given the profound changes brought about by digitization, examining a set of seemingly basic, mundane, and decidedly "non-digital" activities such as food preparation and delivery can help illustrate the many layers of this systemic shift. Preparing and delivering food are hardly new activities, and at first sight they may not seem like intuitive exemplars of the digital economy. In fact, food could be considered a quintessential "non-digital" commodity—given its perishable, sensorial, and nutritional attributes, food cannot (at least at the time of writing) be experienced in digital form, and its consumption generally must take place within a short time from preparation. Even with enormous advances in food transportation and preservation in the past century,[4] freshly prepared food continues to be an enormously popular good, as demonstrated by the billions of people the world over who flock every day to restaurants, eateries, and all manner of establishments where they consume food prepared on demand. Furthermore, food production, distribution, and consumption are deeply rooted in culturally specific and highly localized contexts and thoroughly embedded in all aspects of the economy—both as industries in themselves (e.g., agriculture, grocery, restaurant industries) and as a basic necessity for human sustenance—thus they are fundamental to all other economic activities. It is precisely due to both its local rootedness and its ubiquity that we can take food to be a microcosm of capitalism's digital transformations, where the continuities and ruptures brought on by widespread digitization coexist and interact.

For instance, well before the arrival of food delivery platforms such as Uber Eats, Grubhub, and Doordash, many cities throughout the world had well-established, highly localized, and differentiated markets for the delivery of prepared food. In many cases those markets functioned (and many continue to do so) by customers contacting restaurants directly and ordering "for delivery." Restaurants would then dispatch the order by foot, bicycle, car, or motorcycle directly to the customers' home, workplace, or some other location. The payment would take place either during the initial order (via phone or, more recently, online) using credit/debit card, or upon delivery using cash (or card, if the delivery person had a credit card imprinter or, more recently, an electronic terminal). In these food-delivery markets the people who transported food orders from door to door were often employed directly by the restaurants, and their income would be supplemented by (or often entirely made up of) tips from customers. In some instances, however, this basic arrangement gave rise to particularly sophisticated systems of intermediaries dedicated to connecting food producers and consumers across space.

The lunchbox delivery system in Mumbai known as *dabbawala* is a fascinating example of how such a model evolved and took root in an era prior to the development of digital technologies.[5] This system began to take form toward the end of the nineteenth century, when increased migration from the countryside to the city, a shortage of prepared food establishments, and a gendered and caste-based division of labor created conditions for its emergence in the context of British colonialism in India. According to anthropologist Gauri Sanjeev Pathak, the roots of this system can be traced back to 1890, when Mahadu Havji Bache, "an immigrant to Mumbai from the Rajgurunagar *taluka* (subdistrict) of the Pune district of Maharashtra, discovered that there was a demand for the delivery of home-cooked lunches amongst Indians working for British administrators. To meet this demand, he recruited relatives from within his *taluka*. The system started with around twenty people but rapidly expanded to include members from neighbouring *talukas*."[6]

Over time, this food delivery system has become thoroughly embedded in the urban life of Mumbai. Today it relies on about five thousand couriers who serve hundreds of thousands of customers. *Dabbawalas* (who are overwhelmingly men) collect prepared food from private homes, where food is cooked (usually by women) and packed in cylindrical stacked lunchboxes (*dabbas*) that keep it warm. In complex urban journeys that include transportation by bicycle, train, handcart, and even by foot, dabbawalas deliver the lunchboxes to customers at their worksites by lunchtime. This routine takes place day in and day out with remarkable punctuality and a negligible rate of error. This is because, beyond being a food delivery system, the dabbawala is also a sophisticated, well-organized, and smoothly executed information processing system, where each lunchbox is marked with a unique code that corresponds to the individual dabbawala, the origin/destination, and recipient. By some accounts, around two hundred thousand meals are delivered daily, with a remarkably low error rate that some estimates put at just one in six million deliveries.[7] In addition to processing information in a way that allows for the collection, transportation, and successful delivery of meals, it is important to highlight that the dabbawala system is embedded within Mumbai's particular social structure and division of labor. For example, the dabbawalas constitute a relatively homogeneous group belonging to the Maratha caste. The profession inhabits a space where they have earned widespread praise and affection, but it is not commensurate with their economic circumstances, given their low monthly earnings.

This system depends not only on the dabbawalas themselves, who undertake the food delivery, but on the hundreds of thousands of people who prepare and pack the food every day. This task takes place primarily inside private homes and within heteronormative family structures, where it is carried out overwhelmingly by women. Specifically, food preparation in this context is often an expectation for mothers or wives who stay at home while their sons or husbands go to work. Food preparation in the dabbawala system thus works in tandem with both a specific family structure and the broader division of labor in the local economy. As feminist scholars have long argued, much of the labor needed to maintain and reproduce society is precisely the kind of labor that is both essential to the workings of capitalism while at the same time being generally left out of the formal economy.[8] In the case of the dabbawala system, the labor that goes into food preparation is uncompensated labor that takes place in the household and is mostly done by women. In contrast, the delivery work carried out by the dabbawalas themselves, a mostly male-driven workforce, is part of the market economy and as such compensated in monetary terms. Addressing this tension, feminist media studies scholar Kylie Jarrett argues that "labor done in the home—"women's work"—is a necessary input to capitalist circuits of exchange, producing healthy, socially adept, well-nourished laboring bodies."[9] Yet, despite being essential in many ways, this "feminized" domestic work is often excluded from the monetary sphere and rendered invisible in how we conceive of, and operate, the larger economy. In this case, the food preparation done in the homes is an integral part of the dabbawala system, but is not remunerated with a wage, which can lead some to consider it an entirely separate activity.

In addition to relying on the unpaid labor of preparing food in the home, much of the success of the dabbawala can be attributed to the refinement of an efficient and well-coordinated system of informational organization and delivery logistics. Undoubtedly this system performs an important social function that both employs thousands of dabbawalas and provides nourishment to thousands of Mumbai's workers across industries throughout their workday. However, to understand this system's success it is important to also understand its embeddedness in a particular geographic context and prevailing social order, both of which are heavily shaped by structures such as caste, gender, family relations, and the historical legacy of British colonial rule. In addition to its organizational and logistical innovations, the dabbawala's reliance on the division of labor within families, and between different social groups, has lent stability to this food delivery system and allowed it to become such a fixture in the city life of Mumbai. And yet, while dabbawalas are seen as an esteemed institution by many Mumbaikars, this occupation—and the system it underpins—is under important pressures for change. As in many other cities all around the world, Mumbai has recently seen the rise of food delivery services coordinated via digital platforms. These services simultaneously challenge, overlap with, and extend the existing networks of intermediaries built through systems such as the dabbawala. Considering the transformative forces reshaping food delivery markets in Mumbai (and elsewhere), it is worth examining how these platform-based food delivery networks compare to the dabbawala along social, spatial, and technological dimensions. This comparison can in turn serve as a microcosm where we can glean the emergence and development of digital capitalism, its myriad manifestations across specific geographic contexts, and the spatial architecture that characterizes this new economic formation.

THREE BUILDING BLOCKS OF DIGITAL ECONOMIES

At a surface level the dabbawala system appears to resemble the digital platform–based delivery networks of Uber Eats or Grubhub. After all, in both cases intermediaries are paid to pick up prepared food in one location and deliver it on time in another. However, beneath this veneer of similarity there are significant differences that suggest a broader change in the social, technological, and economic dynamics underlying not only food-related industries and the service sector, but also the broader economy. In this example, and in the rest of the book, I examine how a wide range of activities—such as food preparation and delivery—undergoes transformations through digital technologies, enfolding into digital capitalism and producing new spatialities in the process. To analyze the construction of digital capitalism, and highlight its spatial and geographic dimensions, I consider the three interconnected processes (Figure 1) that together form the core theoretical framework of this book: location, valuation, and marketization (LVM). *Location* identifies how and where things, people, or activities in digital capitalism are situated in space. *Valuation* in turn refers to the assignment of a kind of economic worth (which may be monetary, but not necessarily) to specific elements in the context of digital capitalism, ranging from goods and services to assets, relations, attention, clicks, or myriad others. Lastly, *marketization* describes the construction and workings of the markets that structure digital capitalism, as well as the interaction, circulation, exchange, and regulation

Location

Where are things, people, or
activities situated in space?

Marketization

How are markets
constructed?

Valuation

How do we assign economic
worth to something?

FIGURE 1. Location, valuation, and marketization in the spatial architecture of digital capitalism. Elaborated by the author.

involving constellations of actors, goods, services, and other elements throughout such markets, along with the various outcomes produced by them (e.g., profits, losses, and both negative and positive externalities). Below I explore each of the three processes in the LVM framework as they manifest in the case of the dabbawala food delivery system to illustrate the continuities and transformations brought by the incorporation of digital technologies into various spheres of economic activity as well as social structures and the places and spaces where all of these unfold.

Location

The dabbawala addresses the problem of locating and connecting the sites of food preparation with those of food delivery through an analog information system developed over decades and refined through repeated practice at scale. This system relies on a logically and organizationally coherent syntax of personalized codes physically marked on lunchboxes. This informational system allows for the sorting and allocation of lunchboxes in geographic space with a high degree of reliability. Having accurate and well-organized information, however, is not enough. This is why the dabbawalas' highly refined skill for interpreting and using this information to navigate the urban space of Mumbai is essential for the successful coordination of an enormous number of food deliveries every day. Each dabbawala has developed the practical knowledge of navigating specific routes at particular times and accumulated the social knowledge that comes from interacting with a stable pool of customers day in and day out. In practice this translates to the daily weaving of relatively stable locational networks made up of vast numbers of individual transactions and relations, which in aggregate corresponds to the spatial expression of a citywide food delivery market.

All of this adds up to a very effective system that is nevertheless limited by its very strengths: food is picked up and delivered along the same routes every day, and each route generally connects the same pair of locations in space: a home kitchen with its corresponding office or workspace. Therefore, on the question of location, the dabbawala system is highly effective and specialized but relatively inflexible, since it is tied to very specific social relations, and their corresponding locations (i.e., the same pairs of home kitchens and offices/workspaces). In other words, the advantages that allow dabbawalas to effectively deliver millions of home-prepared meals to office workers do not necessarily translate to changing food delivery contexts—such as customers without someone to cook for them at home or those with unpredictable schedules, who may order food from different restaurants on irregular days.

By contrast, digital food delivery platforms such as Uber Eats, Zomato, and Swiggy (the latter two very popular in India) operate on a different locational logic—one that is largely disconnected from the familial and social relations that support the dabbawala system. Instead, these platforms connect food producers and consumers into continuously reconfigured locational and transactional networks that reflect who orders what, when, and from where at any given time. While some customers may have a favorite restaurant, from which they order with a certain frequency, they are not tied to ordering from said restaurant every day—unlike office workers in the dabbawala system, who receive daily deliveries of food prepared by family members in their own homes. Similarly, even if a customer orders from the same restaurant every day, the shifting labor pool of delivery workers at any given time and place means that there is no intrinsic connection between individual delivery workers and specific locations or the routes that connect them. By the same token, in the food delivery system organized by digital platforms, neither delivery workers nor restaurants (or other food establishments) have a fixed clientele and are instead faced with fluctuations in the volume, composition, and location of orders they receive every day.

This constant reconfiguration of social, spatial, and transactional networks is the result of one of the key features of digital food delivery platforms, which sharply contrasts with the dabbawala system. Platforms like Grubhub, Swiggy, and Zomato operate as intermediaries that connect buyers, sellers, and delivery workers *on demand*. Thus, the connections between pickup and drop-off locations in space are established on a case-by-case and platform-by-platform basis. These connections are established and assigned an economic value through a combination of variables such as user preferences (what people search for), availability (which restaurants are part of the delivery network), hour of the day, and time/distance of each delivery. These and other variables are weighed into the calculations made by the proprietary algorithms of each platform, which ultimately determine not only the price of each delivery (and how it will be allocated among the various parties involved) but also the specific configurations of people and places that assemble into ephemeral networks of buyers, sellers, and delivery workers.

A key aspect of the digital food delivery platform ecosystem (which applies to many segments of digital capitalism) is the constitutive role played by real-time location tracking. The shifting networks woven by connecting customers, establishments, and delivery workers *on demand* are made possible because the location of all parties is tracked in real time by the digital platforms through the GPS capabilities on mobile phones or using IP (internet protocol) geolocation in personal computers.[10] It is this ability to match buyers, sellers, and delivery workers in real space/time, and to provide a means for tracking and communication between them, that underpins the business models of food delivery platforms. While food delivery is not new, and neither is the role of intermediaries, the innovations of digital platforms reside in combining location and other data collection capabilities with an interactive interface that connects and coordinates the various parties to each transaction in real time. However, the change in food preparation and delivery brought by digital platforms is not limited to the remaking of the social and spatial networks involved in each transaction. In other words, digital platforms not only reshuffle who is involved in each food delivery, where they may be located, and how they connect with each other, but they also transform the very nature of what we mean by "food delivery" as an economic activity and the role it plays in market and nonmarket exchanges. A crucial aspect of this transformation is that when transactions are mediated by digital platforms, they involve complex forms of economic valuation that substantively differ from, say, those in traditional food delivery systems like the dabbawalla. These new forms of valuation that arise in the digital economy constitute the second pillar of the spatial architecture of digital capitalism developed throughout this book.

Valuation

As I discussed above, location is crucial to the construction of digital capitalism because it allows one to pinpoint and track particular actors, goods, services, and other elements in physical space—and increasingly in real time. The combination of geospatial technologies (such as GPS) and digital applications (such as mobile digital platforms) produces vast amounts of real-time location data that structure and coordinate networks of buyers, sellers, and delivery workers through food delivery platforms. However, location alone does not fully explain how food delivery (and myriad other activities, products, and services) have become subsumed into the digital economy. To understand this, it is necessary to account for how location is involved in the process of valuation, the second pillar of the LVM framework. Valuation is a highly malleable, multidimensional, and contextual process that can take place in various ways. For instance, location data can itself be endowed with economic value (as is done by firms who act as data brokers), or it can be part of how different actors in the digital ecosystem assign economic value to products, services, interactions, transactions, or relations. Regardless of the specific configuration, location is a key element in many of the business models prevalent in the digital economy. As such, location often enables the very process of valuation at the core of digital capitalism. Over the course of this book, I will explore this relationship between location and valuation across a wide range of examples and show the diversity of their possible configurations and implications. For now, the case of food delivery serves as a useful illustration.

In the case of food delivery (and other services), digital platforms derive much of their economic potential from the fact that they are strategically positioned between buyers, sellers, and delivery workers. Thus, platforms' business models will prioritize translating this intermediary position into monetary value, a process often called "monetization." In a best-case scenario (for the platforms), a successful business model may even lead to actual profits. Often, however, like many of the leading actors in the digital economy, the firms that own and operate digital platforms rely on aggressive growth fueled by outside investments (from venture capital and angel investors, for instance) and operate based on the promise of future profitability resulting from the expectation of a dominant market position. In either case, a central issue for food delivery platforms (and other kinds of digital platforms) is leveraging their position as intermediaries to define who to charge, who to pay, in exchange for what, and how much. As I argued above, location is crucial in this process because it allows platforms to identify the participants in both sides of the market (the customers on the one side, and the restaurants and delivery workers on the other) in real time, at a granular level, and match them according to factors such as proximity and availability. This dynamic and near instantaneous matching process effectively allows a user to place an order from a particular restaurant or establishment and receive it soon after from a delivery worker while primarily interacting directly with a digital interface. Yet, while this process has become so commonplace as to seem straightforward and often seamless, behind the scenes there are multiple forms of economic valuation at work that include, but go far beyond, traditional market transactions, such as direct monetary payments in exchange for products or services.

Through the process of digitization and the role of intermediaries like platforms, what could otherwise be considered a relatively direct exchange of money for food becomes a complex arrangement of commissions, fees, data streams, ratings, analytics, and incentives. These schemes vary across platforms and depend on particular combinations of business models, technological capabilities, and relationships between platforms, restaurants, customers, workers, and regulators. Such variation notwithstanding, a common arrangement can be structured as follows. Customers initiate a transaction by ordering food through the platform, issuing a payment that is split, in the first instance, between the restaurant and the platform itself. This payment can be broken down into the price of the food, a delivery fee, taxes, other fees, and a tip for the delivery worker. Separately, restaurants often pay a certain percentage of each order to access the platform's customer base and pool of delivery workers. The delivery workers, on the other hand, get paid directly by the platforms, although this can be supplemented with tips from customers.

An important aspect of the role of delivery workers, which has been the point of much contention and legal battles throughout the world,[11] is that they are often not considered to be employees of the platform, but rather independent contractors. This employment status in turn helps explain both the power and the economic value that the platforms derive from their position as intermediaries. While the breakdown can vary by company, in general terms, food delivery platforms earn money in the form of commission from the other involved parties (customer and restaurant), and by setting the formula by which delivery workers get paid.[12] However, these terms only cover the basic contours of each transaction. To understand how platforms engage in valuation it is necessary to appreciate how they combine capabilities for real-time monitoring of the levels, fluctuations, and spatial distribution of supply, demand, and delivery workers, as well as providing a mediating user interface where consumers and all other parties interact. An example of how platforms exercise this combination of capabilities is through the mechanism of "surge" or "peak hour" pricing, which increases delivery surcharges and/or other commissions according to real-time supply and demand conditions in particular geographic locations. Another is advertising, through which delivery platforms can charge restaurants for a preferential position in the resulting map display or list that appears when customers use the digital interface. This way, in addition to providing delivery service, platforms can also act as advertising spaces that commercialize visibility within their network of buyers, sellers, and delivery workers.

Even from this cursory look it is evident that the valuation processes that take place within food delivery platforms involve much more than the service of food delivery itself. Identifying such valuation processes in turn sheds light on some of the fundamental economic and political questions underlying digital platforms (e.g., what is being bought, what is being sold, what is being exchanged in nonmonetary ways, who gets what, and under which conditions), bringing into focus new arrangements of social and technological relations. In some ways, these relations resemble those that constitute traditional systems of food delivery, but they also differ in important respects. To illustrate these changing social relations, shifting spatial configurations, and their connection to processes of valuation, it is worth expanding on the comparison between digital food delivery platforms and the dabbawala system of Mumbai.

Both digital food delivery platforms and the dabbawala rely on a workforce of intermediaries that connects the locations where food is prepared to those where it is consumed. However, in the dabbawala the networks of locations are qualitatively different from those in food delivery platforms in at least three respects. First, individual dabbawalas have stable sets of customers, whom they visit repeatedly. Second, the locations where the dabbawalas pick up and deliver food also tend to be stable because they usually constitute pairs of homes and offices. Third, precisely because of the configuration of these networks, customers do not pay for food preparation, since this task is generally carried out at home by family members. However, customers do pay the dabbawalas directly for the delivery service. Thus, digital food delivery platforms and the dabbawalas do not only operate across space in different ways, but they differ more fundamentally in their very spatial constitution.

Examining the differences in the networks of locations, transactions, and social relations that emerge through both the dabbawala and digital food delivery platforms highlights the distinct logics of valuation underpinning these two systems. At one level, these logics are manifest in the various arrangements of transactions, payment formulas, and business models described above. Such logics are also informed by the relative positions of key actors between each other and within each system. For instance, platforms on the one hand, and dabbawalas, on the other, take on different roles within their respective ecosystem, each mediating information flows, structuring exchanges, and navigating or enacting power relations between other actors such as customers, food preparers in private homes, restaurants, and advertisers. In aggregate, a crucial difference between these two systems is how they are embedded into, or integrate with the specific social, political, and geographic configuration of each place where they operate.

While the dabbawala system relies heavily on stable social structures shaped by the specific caste, gender, and class dynamics of Mumbai's urban context, digital food delivery platforms still rely on such localized dynamics, albeit in less rigid ways—sometimes exploiting and sometimes remaking them according to the needs and priorities of the platforms themselves. Yet, even as digital platforms might appear to be less embedded in the social relations of each place (in contrast to the deep embeddedness of the dabbawala system), it is important to emphasize that the logics of valuation in each of these systems can never operate in a vacuum, for they always require physical and social space to unfold and are thoroughly structured through spatial logics. It is therefore essential to explain how digital platforms, or traditional food delivery systems relate to larger structures of exchange and broader social and geographic contexts. To do this, it is necessary to explore how markets are assembled (and, specifically, spatially constituted) in ways that correspond with particular contexts. In this book I propose to examine this diverse configuration of markets through the lens of *marketization*, the third pillar in the LVM conceptual framework underpinning the spatial architecture of digital capitalism.

Marketization

Food delivery systems, from those we might consider "traditional" such as the dabbawala, to those enabled by digital platforms such as Grubhub or Swiggy, are always part of larger structures of exchange, which themselves operate within broader social and geographical contexts. These structures of exchange are markets, which are a central feature to capitalism, though not necessarily exclusive to this economic system. Regardless of where they are found, however, or the system within which they operate, markets are always spatially constituted. This means not only that markets are located in space, but that space plays a substantive and active role in their configuration and dynamics, far beyond merely acting as a backdrop or a container. If we understand markets this way, we should expect them to show a great deal of differentiation from place to place due to their close integration with a range of contextual and geographic factors such as proximity, scale, local culture, institutions, and territory. This is a departure from dominant perspectives, such as those that have emerged out of mainstream economics, which tend to analyze markets as relatively idealized formations that are often highly abstracted from the specific social or geographic contexts where they operate. In this view, markets behave according to a relatively predictable set of laws and generally trend toward equilibrium through processes like market clearing, where the supply is equal to the demand, and no quantity is left over. By contrast, in geographical studies of markets, the attention shifts away from general laws or rules that market dynamics are expected to follow— for example, there is no reason to expect markets to trend toward equilibrium.[13] Instead, understanding markets geographically privileges their difference, contingency, variation, and embeddedness within specific spatial and social contexts.[14] This emphasis on the geographic dimensions of markets becomes even more necessary when dealing with goods and services that have often been labeled "immaterial," such as those that make up the digital economy. Emphasizing the geographic dimensions that shape markets for "immaterial" goods is necessary to counter a prevalent tendency to think of such markets as placeless, and by extension more difficult to pin down, manage, or regulate.

In more general terms, whenever the geography of markets is mentioned, it is often done so in reference to factors closely linked to the physical environment, such as natural resource endowments (oil, timber, corn, etc.) or the costs of transporting inputs and goods in the process of economic production, distribution, and consumption. However, while a geographic approach to markets indeed considers such factors, it also encompasses a much broader and deeper set of considerations that, in aggregate, fundamentally alter the very notion of how we conceive of markets and how they operate in the world. A useful shorthand is that a geographical perspective can help us understand not only where markets are located *in* space and why, but also how they are constituted *by* and *through* it. This approach becomes especially useful when the goal is to study digital information, which has often been erroneously assigned to a realm outside of geographic space and even beyond the material world altogether.

Notions such as "cyberspace," "the cloud," and even the names for technologies like "ethernet" and "virtual reality" suggest that digital information is somehow intangible or devoid of physical manifestations. A consequence of this view is the misunderstanding that flows of digital information can be "everywhere" and "nowhere" at the same time. In other words, in this interpretation digital goods and services float free from spatial constraints, unaffected by geographic factors that may impact markets for other goods whose materiality is not up for debate (for instance, brick-and-mortar retail, or agricultural products). The aspatiality inherent in such notions in turn helps explain the recurring allure of predictions that digital communications will "collapse space," spell "the death of distance," or precipitate "the end of geography."[15] The arguments developed throughout this book seek not only to dispel such notions of "the digital" as largely aspatial, immaterial, or somehow outside of geography, but to go further by demonstrating that digital capitalism itself is a thoroughly and inevitably spatial project and by showing how its geographic dimension matters a great deal for its configuration, dynamics, politics, and outcomes.

To understand markets from a geographic perspective it is necessary to account for how they come together in ways that may sharply diverge from theoretical constructions that privilege general laws and generalizable outcomes. Thus, geographies of markets pay attention to the factors that shape such divergences from the idealized model of "the market" and emphasize their spatial variation and distinctiveness while attending to their potential commonalities. These factors include the degree and kind of urbanization of a particular place, how it may be connected to other places, the locally specific configurations of cultural forces, political dynamics, the dominant legal, regulatory, technological, and infrastructural conditions, how these conditions relate to processes of production, distribution, and consumption in a market, as well as the strategies and practices of firms, consumers, and other actors such as the state. Prioritizing how these and other factors come together across different contexts is meant to illuminate, in all their "messiness," how markets work in the real world, rather than abstract them into idealized models from which every real instance is a deviation of some kind. This is not to say that model-based approaches are not useful to understand certain aspects of markets in the real world. Conversely, there is no necessary reason to think that geographic perspectives of markets are incompatible with quantification or abstraction. Rather, the point of departure between the two approaches is one where spatial variation and context are seen as intrinsic (and indeed fundamental) to market formation, leading to a different set of possibilities of what markets can be, how they function, how they can be studied, the outcomes they may produce, and how they can be managed and regulated.

As I discussed above, location and valuation can help elucidate how digital goods and services are connected to geographic space and endowed with different forms of economic value. Bringing in marketization, or the process of market creation, as the third element of the spatial architecture of digital capitalism can then help answer the question of how such digital goods and services circulate in structures of exchange, and how such structures are operated, maintained, and regulated in different contexts. Throughout this book I will pay attention to elements like the infrastructural dimension of digital communications, jurisdictions and legal regimes, and the practices and expectations of user communities and other stakeholders. However, the broader point is that a geographic perspective of digital markets connects specific digital goods and services with the structures of exchange through which they circulate, and in turn situates these structures in concrete places and contexts, transforming such places and contexts in the process. Together, LVM, representing the overarching integration of location, valuation, and marketization then gives us a synoptic perspective of how digital capitalism is underpinned by a particular spatial architecture that, though varying from place to place, follows an overarching logic of systemic coherence.

Returning to the example developed throughout this chapter, as recounted above the dabbawala system has deep roots in Mumbai's history, social structures, and local urban context, while simultaneously operating within the city's broader markets for food, groceries, and delivery services. While digital food delivery platforms may lack such deep roots, this does not mean that they are divorced from local contexts and markets in the places where they operate, but rather that their integration has a different configuration—one that may not be as immediately evident. Understanding markets as spatially constituted also means taking seriously their relationality. In the case of the two food delivery systems under discussion here, this means not only acknowledging that they are integrated into the contexts of specific places (albeit in different ways), but also that they are engaged in continuous interaction and reciprocal transformation.

Social and technological changes can in turn reconfigure the division of labor, the locations for food preparation, and the linkages with the locations of consumption, resulting in new arrangements that can circumvent demand for the dabbawala while creating a space for more flexible delivery services, such as those provided via digital food delivery platforms. While this particular reconfiguration of the local food delivery market can be partially spurred by demographic changes (such as the rise in unmarried men), other types of changes in the local context may compound it. For instance, a transformation in the social norms about gender may impact the division of labor, decreasing the expectation that married women prepare the food at home, and opening the possibility that they may occupy their time with other activities (such as their own personal interests or professional development). Yet another force for change may be the availability of a flexible labor force that is captured by digital platforms, which can in turn provide services that were previously only available to a segment of the population, and in a particularly rigid social and spatial configuration (e.g., the home-office link of production and consumption discussed above).

This is an example of how location, context, social structure, and other place-based factors can converge to shape both traditional and digital goods and services. Conversely, as in the case of the dabbawala and the digital food delivery platforms, changes in different modes of production, distribution, and consumption are continuously interacting, engaging in feedback loops and, in turn, reshaping the very configuration of places themselves. For this reason, it is essential to understand the emergence and dynamics of markets in digital capitalism, not as an undifferentiated, placeless (and weightless) phenomenon, but rather as combinations of technological changes, social forces, and political economic conditions that unfold differently in specific geographic contexts. As suggested by the LVM framework, to analyze the configuration of such markets and their geographic dimensions, it helps to learn how digital information is pinpointed to specific locations in space, and how this process is fundamental to assigning value to digital goods and services in different ways. Collectively, location, valuation, and marketization can help us understand how digital capitalism is far from abstracted from physical reality or disembodied in any significant way. To the contrary, adopting a geographic perspective, and specifically the LVM framework, reveals how digital capitalism is thoroughly localized, embedded in specific places, and made and remade through spatial processes—which in turns means that it can be analyzed, contested, and regulated by leveraging specific tools, institutions, and mechanisms that are themselves spatially constituted across scales, from the local and regional, to the national, and international. Taking these arguments as a starting point, the remainder of the book will follow a series of threads interweaving digital capitalism with its many geographies, highlighting the role of geographic information, media, and technologies. Prior to that, however, the discussions in the following chapters must first be contextualized in the emerging body of scholarship aimed at studying digital geographies.

THE DIGITAL IS GEOGRAPHIC

In the following paragraphs I make the case that it is necessary to understand the system of digital capitalism in geographic and spatial (as opposed to only or primarily technological or economic) terms.[16] This argument is at the core of the book's theoretical approach, which is developed by examining the technological capabilities, processes, and outcomes of linking digital information to specific geographic locations and social contexts. The focus is on spatial tools such as GPS, IP address geolocation, and the sprawling digital mapping ecosystem known as the *geoweb*, which encompasses different kinds of geographic information on the internet, from services like Google Maps to geotagged and georeferenced social media content.[17] A key premise of the book is that such geospatial technologies do more than simply provide geographic information in digital form; rather, they transform space and spatial relations in the process, leading to new and different economic and political formations, such as those that characterize digital capitalism. For the purposes of understanding digital capitalism in spatial terms, a key step is documenting how digital geospatial technologies form a connective tissue that brings the transactions and interactions of the digital economy into the physical and social reality of places, spaces, and territories. As I show throughout the book, in doing so, geospatial technologies are instrumental for *location*, *valuation*, and *marketization*, processes central to the spatial architecture of digital capitalism.

The view of capitalism I adopt throughout the following pages is that of a highly differentiated and adaptable social, technological, and political-economic system that is both extraordinarily variegated in its outcomes and contexts while remaining, at its root, a coherent project characterized by a set of shared underlying logics. While in the next chapters I emphasize the role of digital technologies in enabling *digital* capitalism, my examination of this system is informed by a body of work that is not exclusively centered on digital technologies. Distributed across the social sciences (from geography to political economy and sociology), the scholarship on *variegated capitalism* provides a helpful framework that identifies the common threads throughout capitalism's myriad geographic variants while acknowledging its complex interdependencies with local contexts and place-specific factors. This attention to capitalism's simultaneous plasticity and its underlying consistency provides a useful lens to examine the commonalities and differences in processes and outcomes found across locations and scales.[18] Building on this foundation, the arguments in this book take on the task of explaining some of the key dynamics of digital capitalism and the geographic logic that underpins them. To do this I draw from a robust body of scholarship that has, for the past two decades, investigated the geographic dimensions and implications of the emergence and spread of digital technologies.

The development of information and communication technologies (ICTs) throughout the twentieth century was characterized by increasing speed and coverage, bringing television, telephones, and home computers to billions of households around the world. The digital revolution started with workplace, and then personal, computing, and was further accelerated when the internet was opened for public access and dramatically expanded in coverage throughout the second half of the 1990s. This created a qualitatively different way of accessing information, experiencing space, and connecting places to each other. In geographical scholarship these concerns shaped debates addressing the idea of time-space compression, articulated by David Harvey in his 1989 book *The Condition of Postmodernity*. Reflecting on Marx's notion of the "annihilation of space by time," Harvey linked the shrinking times and distances between places that resulted from advances in communication and transportation with the underlying dynamics of an increasingly globalized capitalism.[19] In the resulting "time-space compression," space is rearticulated according to the demands of the new post-Fordist (and postmodern) capitalist economy, such as emerging forms of flexible production and distribution. Among the many impacts of this rearticulation, scales that are central to the spatial logic of capital (such as the global) can override and redefine the priorities and rhythms of those that are secondary (such as the local).

The experience of time-space compression, however, is not solely determined by the mechanics of global capital. Eloquently illustrating this point in her memorable essay "A Global Sense of Place," Doreen Massey argued that time-space compression is not experienced by all places and all people equally or evenly but is instead characterized by a *power geometry*, "for different social groups, and different individuals, are placed in very distinct ways in relation to these flows and interconnections."[20] In other words, we must attend to the characteristics of not only location, but class, race, gender, and other social and spatial axes of difference that shape who has power over, or benefits from, space-time compression, and who bears the brunt of its impacts.

By the early years of this century, the extent of the industrial, societal, and spatial transformations wrought by digital technologies was visible in many aspects of daily life, but the underlying mechanisms were still scarcely understood. In this context, research from geographers and urban planners such as Matthew Zook, Yuko Aoyama, Martin Dodge, and Anthony Townsend contributed to develop fuller explanations of these changes through detailed accounts of the internet, the rise of the information technology industry, and the impacts of this communications network in a variety of productive sectors—from telephony to advertising, media, and retail—all of which resulted in the proliferation of new "internet geographies."[21]

With the economic and cultural ascendance of cities at the turn of the century came the accelerated entwining of digital technologies in urban life. In this context, geographers turned their analysis to how such technologies were reshaping not only how cities work, but actively redefining the very fabric of space. Stephen Graham and Simon Marvin, for instance, studied the ever more important, yet often invisible role of telecommunications networks in making cities function while connecting them with each other. In doing so, they highlighted how infrastructure is inseparable from political economic struggles at the urban and other scales. Furthermore, they also showed how, in an era characterized by the expansion of markets and the retreat of public goods, the digitization of cities has led to a "splintered" form of urbanism. Consequently, the software-sorted geographies that resulted from wiring cities through digital networks, while enabling new services and affordances, have also facilitated the enhancement of surveillance, control, and militarization of urban life.[22] At a fundamental level, these transformations can be understood as part of the myriad intricate ways in which the work of software code reshapes the construction, experience, and functions of space and the social relations around it, leading to the emergence of what Rob Kitchin and Martin Dodge have termed *code/space*.[23]

Beyond the spatial reconstitutions of the urban fabric associated with digital technologies, geographical research has also explored the patterns and determinants of diffusion and distribution of digital communications networks.[24] Several scholars have shown that digital networks and related technological innovation patterns do not appear on a blank slate but build upon previous rounds of infrastructural and economic development, from telegraph networks to railway lines—exhibiting high levels of urban and developmental path dependencies.[25] Simultaneously, the expansion of networked technologies has also articulated an ever-finer and increasingly complex global division of labor. Connected by digital networks, this global division of labor has reshaped economic activity through increasingly digitized production techniques, distribution processes, and marketing tactics. Accordingly, these changes are associated with new categories of goods and services, created, or transformed through digital technologies (ranging from physical goods like cars and consumer electronics to fully digital goods, from e-books to streaming multimedia content). These "digitized" goods and services are in turn accompanied by new patterns and practices of consumption (such as online shopping, mobile applications, crypto-currency transactions, and even virtual purchases within video games). However, as the economy has become thoroughly intertwined with digital technologies, it has become characterized by two seemingly contradictory tendencies: increases in spatial agglomeration (cities have become even more important centers of the world economy) have been accompanied by the accelerated spatial disaggregation of ever-increasing economic activities (for instance, through the ubiquity of value chains stretched across the globe).[26] In other words, digitization and agglomeration have together produced simultaneously centrifugal and centripetal effects on the map of the world economy.

Such seemingly paradoxical geographies point to a fundamental change in the very constitution of the capitalist economic system; a change catalyzed not only by the expansion of digital networks but also by the undeniable import of digital information across realms of economic activity. Aiming to shed light on this shift, economic geographers have spent much time and effort over the past two decades documenting the layered impacts of an increasingly digitized and interconnected global capitalist economy. Chief among these are the augmented complexity and salience of knowledge generation and circulation. This has augmented cities' already central place in the global ecology of knowledge. Owing to the forces of spatial agglomeration and their role in creating sites for dense networks of social, material, and informational exchange, cities have long been humanity's centers of creativity and innovation.[27] However, in the past few decades, as knowledge became simultaneously more complex, varied, and specialized, the geography of its production and its integration into capitalist economic activities (for example, through patenting and other forms of innovation) became increasingly reflected in patterns of intensified spatial clustering.[28]

The result is a geography articulated by a network of city-regions that act as global centers of specialization in such industries like information technology (Silicon Valley, Bangalore), film and television (Hollywood, Bollywood, Nollywood), fashion (Paris, Milan), and finance (New York, London, Hong Kong). This geography of localized specialization, global interconnection, and pervasive digitization has both arisen from, and further cemented, a qualitative shift in capitalism. Underlying this shift is an expansion in both supply and demand of goods and services defined by their salient creative, cultural, cognitive, informational, and symbolic attributes.[29] Thus, the capitalist economy of the twenty-first century is shaped by the rise of industries such as finance, education, health care, media and entertainment, information technologies, and myriad personal services. This economic landscape is in continuous transformation, not least due to the effects of digitization. Thus, as new industries acquire greater salience, traditional ones are reinvented. In this way, "traditional" activities such as manufacturing, agriculture, and energy extraction are transformed into increasingly differentiated, information-intensive, digitized, and often design-oriented, craft-like, and niche-creating industries. In this context, the expansion of "born digital" goods and services such as app stores and other digital marketplaces, video games, online streaming, social media celebrities and "influencers," and internet advertising goes hand in hand with the popularity of traditional ones that have been reimagined, such as mass-customized goods like made-to-order shoes, personalized mattresses, and specialty agricultural, grocery, and gastronomic offerings, such as subscription coffee beans and tropical fruit boxes or home-delivery meal kits.

Attending to these fundamental changes at the core of capitalism, geographers and scholars in cognate fields have characterized the resulting economy, variously, by emphasizing its cognitive-cultural orientation, documenting its thorough reliance on digital technologies writ large, or honing in on the increased importance of digital platforms, their data collection logics, and business models.[30] Documenting this digitally mediated reconfiguration of capitalism, throughout the past decade geographers and urbanists have studied disruptions and transformations in many of the industries at the core of this new economy such as film, television, fashion, comic book publishing, animation, video games, and the performing arts, to name a few.[31] While this rich body of work tells us much about how the digital economy is intertwined with myriad technological, cultural, economic, and indeed geographic forces, I contend that the latter aspect of this relationship demands greater elaboration. While scholars have indeed attended to the various geographic aspects of the emerging digital economy, less attention has been paid to disentangling the spatial logic undergirding the very digitization of capitalism. This logic certainly relates to the spatial impacts and expressions of the digital economy, such as the "internet geographies" alluded to earlier. Yet it points to something more fundamental, as it lies at the heart of the very conception of space embedded in digital technologies.

In other words, to understand how "the digital" is transforming the capitalist space economy, we need to understand not only how it connects and transforms places, or how it disrupts and remakes industries, but also how it encodes, represents, reproduces, mediates, and even re-creates space in fundamentally different ways to prior technological paradigms. The goal of this book, then, is to produce a unified framework to understand the spatial logic and the underlying architecture of digital capitalism. The basis of this framework rests on the foundational role played by geographic information technologies in the digitization of location (and other spatial features), in its valuation in economic (and other) terms, and in the construction of digital markets. I argue that this triad of processes, shortened as LVM, and enabled by different forms of geographic information, media, and technologies, is fundamental to the construction of digital capitalism, and therefore to the capitalist space economy in the twenty-first century. For this reason, building the LVM framework requires integrating an economic geographical perspective with accounts of the expanding universe of geographic information and digital geospatial data on the internet and beyond.

The Geographic Becomes Digital

As Rob Kitchin has identified, the data revolution is characterized by dramatic increases in the volume, variety, and velocity of data, all of which have been enhanced by digitization.[32] One of the effects of this revolution has been the growing presence and circulation of geographic information in digital networks, particularly the internet. While geographic information systems have existed since the 1960s, first in analog, and then in digital form, the development of geospatial technologies has benefitted dramatically from the widespread use of digital computers and networks. Thus, today digital services and platforms from Google Maps to Uber, Yelp, Facebook, Twitter (now X), Amazon, Netflix, and countless others incorporate enormous amounts of geographic information, mapping applications, new spatial media, and locational practices. This integration of geographic and spatial data into the internet is often referred to as the *geoweb*, or *geospatial web*, and considered a distinct and innovative informational environment in its own right.[33] Among myriad others, the geoweb encompasses applications such as open government spatial data portals, commercial mapping platforms, user-generated geo-tagged and geolocated content, citizen science projects, and targeted advertising, many of which rely on (different kinds of) precisely captured locational data supplied by users, knowingly or unknowingly.

Storing and disseminating geographic information on the internet has been transformative in many ways. By combining vast spatial data archives with user-facing portals and information management tools, services in the geoweb can both enrich spatial data and make it more easily shareable by states, cities, corporations, nonprofit organizations, and the public. The shareability, interactivity, and increased role of users in the production and dissemination of data made the geoweb a prominent example of a transformation that took place on the internet during the first decade of the twenty-first century. This shift, dubbed the rise of "Web 2.0,"[34] referred to the appearance of more interactive, mobile, and social digital services that remade the internet, which had until then been a landscape characterized by relatively static pages and sites. The appearance and widespread popularity of social media platforms such as Facebook, Twitter (now X), Instagram, (Sina) Weibo, WhatsApp, and VKontakte then potentiated the relational and user-driven aspect of the internet during the 2010s. While social media itself integrates vast amounts of geospatial data (from addresses to geotagged content), the geoweb also grew thanks to dedicated geospatial data communities (like OpenStreetMap), new forms of navigation (like Google Maps and Waze) and the normalized use of location and other forms of geographic data for activities as diverse as advertising, ride hailing, and the delivery of digital and physical goods and services.

While the geoweb can be considered a subset of a broader digital information environment, geographers and other scholars have argued for and studied its unique dynamics. A salient component of the geoweb is the rising prominence of *volunteered geographic information*, or VGI. This kind of geographic information encompasses collaborative citizen science projects like iNaturalist, a joint initiative of the California Academy of Sciences and National Geographic Society, where participants, acting as citizen scientists submit georeferenced and tagged observations of biodiversity that are then mapped throughout the world. Another popular project is OpenStreetMap, where users can contribute names, traces, and other geographic information to a worldwide interactive, open, and collaborative mapping project and geographic database. However, VGI can also refer to a broader category, such as the geotags and other forms of geographic identifiers submitted by users when sharing text, images, videos, or other media via social networks like Facebook, Instagram, TikTok or Twitter (now X). Research focusing on VGI has emphasized the participatory networks that lead to the creation and dissemination of vast amounts of information in the geoweb by nonprofessional user communities. Among other qualities, VGI has been considered to have high scientific and societal value due to three key qualities: timeliness, low cost, and extensive coverage.[35] Scholars studying VGI have shown the wide range of outcomes and impacts of crowdsourced geographic information: from producing geographic knowledge to leveraging public participation for scientific and government projects to addressing complex environmental and societal challenges, such as severe weather events and public health crises.[36] While there remain important challenges in the production and implementation of VGI, such as privacy concerns, exploitative practices in the commercialization of personal information, and significant variations in data quality, this type of geographic information represents a change in the relations of production of a kind of information that has traditionally been the exclusive domain of state-funded agencies. Hence, VGI also embodies a specific subset of the many possible avenues for participatory engagement in the geoweb.

The participatory nature of much of the geoweb, along with the range of applications, from scientific to humanitarian, produced a sense of optimism that centered on its potential to produce tangible social and environmental benefits by leveraging the innovative and disruptive qualities found in new forms of producing, distributing, and even imagining, geographic information. This wave of optimism also brought forth arguments advocating for the role of the geoweb in democratizing information, contesting traditional hierarchies of expertise, and challenging established epistemologies of geographic knowledge.[37] Throughout the mid-2000s, some argued that the expansion of participation enabled by the geoweb, and its myriad technical, political, and epistemological ramifications had brought about a new kind of geography, which was termed *neogeography*.[38] While acknowledging the importance of the geoweb, and digital geographic information more broadly, geographers set out to temper some of these claims, situating neogeography in a broader context of changes in the production and distribution of geographic information. This in turn contributed to nuancing and contextualizing not only the very real possibilities and potential benefits of the geoweb, but also to highlighting its limitations, challenges, and risks, as well as its relationships to existing systems of social, political, and technological organization.[39]

As evidenced by the enormous popularity of applications such as Google Maps and OpenStreetMaps; Yelp; TripAdvisor; platforms like Uber, Airbnb, and Instacart; and even open government data portals, widespread use of location and other forms of geospatial data on the internet has played a crucial role in structuring today's digital environment. However, each of these and myriad other cases also exemplify why we should also pay attention to how geographic information on the internet (in other words, the geoweb) relates to broader social forces and political economic structures—from data commercialization to government and corporate surveillance to transparency and regulation. An important point to highlight in this respect is that the distribution of mapping tools does not necessarily imply their successful use or adoption.[40] As geographer Muki Haklay has cautioned, we should be careful with how we interpret the widespread popularity of the geoweb, particularly with respect to its political implications. Specifically, we must avoid conflating the broader distribution of technical tools and participatory platforms with processes of political democratization.[41] While these processes are not necessarily in contradiction with each other, neither should they be seen as mutually reinforcing. In the past decade and a half, public discourse has alternated between these positions, first with enthusiasm after the Arab Spring—often using monikers like Twitter (now X) Revolutions—to more recent concerns stemming from society's reliance on digital information, such as the propagation of misinformation and disinformation through social networks, as well as its political consequences, especially in the aftermath of the 2016 and 2020 US presidential elections or the accumulation of power and market control by tech conglomerates, as evidenced by the rising discussions surrounding antitrust regulation in the United States and other jurisdictions.[42] As these overlapping and often clashing debates suggest, the widespread availability of digitized information cannot be equated with any single societal outcome, and in fact can have multiple political valences depending on specific contexts and situations. More specifically, while the geoweb has embodied many benefits from participation, collaboration, and new ways of representing, and interacting with space, it has also shed light on the urgency of a number of issues stemming from the widespread use of location and other geospatial data, from privacy regimes to data commodification and commercialization to cybersecurity, persistent inequalities in information access, and the massive use of geographic information for surveillance by state, corporate, and other actors alike.[43]

The development of the geoweb, and discussions surrounding it, are closely related to the "big data" coverage and debates during the 2010s. As suggested above, geographer Rob Kitchin's succinct characterization of volume, velocity, variety, reminds us that although large datasets are not new, the new digital environment does create a particular set of conditions that should motivate a reconceptualization of "big data."[44] Such large datasets, which increasingly circulate on digital networks and include (among others) volunteered, extracted, and automatically generated digital information, are today fundamental to the operations of governments, private corporations, researchers, and citizens alike—albeit characterized by vast asymmetries in technical capabilities and political and economic power.[45] In attending to the issues brought about by big data, and specifically big geospatial data—such as that circulating in the geoweb—geographers and other scholars have contributed to the development of critical data studies, an approach that enables us to understand and document how data are intertwined with other forces that make up our social world such as changes in capitalism, the emergence of new social norms, processes of innovation, and institutional transformations.[46]

The conversations surrounding critical data studies and other examinations situating the geoweb in its social, political, and economic context can be linked to prior debates that gave rise to subfields such as "Critical GIS" (geographic information systems) and "critical cartography" throughout the past three decades.[47] Both of these respond to the need to take seriously the power dynamics that both enable and emerge from GIS, maps, and other spatial technologies. Also known more broadly as "GIS and society" debates, collectively these conversations demonstrated that maps and geographic information more generally cannot be understood while ignoring power relations (between states, firms, people, and other actors).[48] Similarly, maps—analog, digital, or otherwise—cannot be taken at face value as impartial, unfailingly reliable representations of the world. Quite the contrary, maps and geographic information are inherently situated, partial, and selective views of reality that should in turn be understood as part of broader social processes. This means that the intersection of axes of differentiation and domination, from gender, class, and race to ideology, political economy, cognitive biases, and technological affordances, are as important to understanding maps and geographic information as technical features like projections, color gradients, elevation curves, and symbolization. Building on these arguments, critical data studies and digital geographies have expanded conversations developed in the past decades by scholars, work in the critical, participatory, and feminist GIS traditions, producing a rich and diverse set of tools, theories, and methods that can assist us in understanding how geographic information both helps shape, and is irremediably influenced by our social and political world.[49]

As geographic information systems have gone from analog to digital, and now to networked within the geoweb, these technologies have catalyzed multiple innovations in cartography, spatial data, and forms of mapping and viewing the world. In all of this, geographic information has become more accessible and interactive, shaped by widespread participation and new practices of geographic knowledge production. At the same time, new challenges and perils, from privacy and cybersecurity to exploitation and surveillance, shape the landscape of the geoweb. Both of these developments highlight how the geoweb, and geographic information, more generally, are intertwined with specific configurations of technologies, social practices, political economic processes, and legal and regulatory frameworks.[50] For this reason, it is necessary to bring together discussions on critical data studies of the geoweb with a broader political economy that examines the construction and development of digital capitalism, and particularly the role of geographic information in this process. It is this question that motivates the examination of the spatial architecture of digital capitalism at the backbone of this book.

THE SPATIAL ARCHITECTURE OF DIGITAL CAPITALISM

The arguments of this book are developed through a series of case studies of industries, platforms, and technologies where digital geographic information plays a prominent (if not always evident) function. The next chapter examines the emergence and evolution of MapQuest and Google Street View, two key components of the online mapping ecosystem known as the *geoweb*, some of which prompted the discussion above. The third chapter addresses the link between location, geolocation, and allocation. The focus is specifically on showing how technologies like GPS and other geolocation tools have been incorporated into the internet, primarily for the purposes of targeted advertising, geographically tailored content delivery, and the construction of digital markets. Chapter 4 documents the emergence of a new (outer) space economy bolstered by privatization of space travel and the development of new miniature satellites for Earth observation imagery and telecommunications. Chapter 5 follows the development of two new digitally enabled disruptions in the system of automobility: the development of autonomous vehicles and the rise of ride-hailing platforms. Each of these chapters addresses the crucial role played by different forms of geographic information in underpinning digital capitalism and its spatial architecture. The logic underlying this spatial architecture is analyzed throughout these case studies through the three conceptual key pillars that constitute the LVM framework: *location*, *valuation*, and *marketization*.

CHAPTER 2

Assembling the Base Map

From MapQuest to Google Street View

INTRODUCTION: THE EVOLVING LANDSCAPE OF DIGITAL MAPS

This chapter addresses the emergence and transformation of the on-line digital mapping ecosystem throughout the first two decades of the twenty-first century. It begins by describing early successful web mapping ventures such as MapQuest, which provided static web maps with routing directions for drivers. MapQuest marked a turning point both in the geospatial data economy and in the development of the internet more broadly because it drastically opened access to personalized digital maps, which had previously been only available to researchers, professional cartographers, and other users with access to specialized tools like geographic information systems (GIS). To explain the significance of MapQuest, the chapter discusses some of the key factors that made this platform possible—and with it, laid an important building block for the geospatial data economy as we know it today. In particular, the chapter highlights the development of GIS for computer mapping in the 1960s and 1970s in the United States and Canada; the availability of satellite imagery, particularly from NASA's Landsat program; and the creation and distribution of TIGER files (Topologically Integrated Geographic Encoding and Referencing; see Figure 2), with which the US Census Bureau pioneered linking geographic data (such as county boundaries) to demographic variables—and eventually countless other data categories, which would pave the way to the geodemographics industry.

FIGURE 2. Logo for the United States Census Bureau's TIGER (Topologically Integrated Geographic Encoding and Referencing) map data format. *Source:* US Census Bureau.

Understanding these technological developments allows us to not only explain the rise (and eventual decline) of MapQuest itself, but more broadly to shed light on this company's influence in the development of the geoweb, an entire ecosystem premised on the incorporation of geospatial information on the internet at the turn of the twenty-first century. The chapter then discusses the subsequent expansion and transformation of the geoweb through the entry of Google into this online mapping ecosystem. While Google's geospatial turn was a sign of the broader salience of geographic information on the internet, and web mapping in particular, this event was particularly significant because it led to the creation of standard-setting services like Google Maps, Google Earth, and then Google Street View. Google's geospatial services increased both the accessibility of online maps and their geographic coverage while transforming user experience through interactive and navigational features. These developments did not take place in a vacuum, but in a growing landscape of online geospatial services, platforms, and portals, populated by competitors like Bing Maps, Apple Maps, national and subnational government data portals, crowdsourced platforms for mapping (OpenStreetMap), and georeferenced street-level imagery (Mapillary, KartaView). As discussed earlier, the explosion of mapping products and services on the internet catalyzed the formation of a new environment for geospatial information online, which came to be known as the "geoweb." This environment was characterized by quantitative increase, diversification, and qualitative shifts in the production, circulation, and applications of geospatial information on the internet throughout the 2010s. The geoweb was not limited to mapping platforms, and it would eventually encompass a vast universe of geographic and location data (from GPS coordinates to geotagged photographs and geolocated tweets). Yet, despite the informational diversity of the geoweb, focusing on the development of key online mapping platforms and tracing their broader ramifications can reveal some of the core logics through which geographic information has come to acquire such a prominent role in digital capitalism.

While early entrants like MapQuest pioneered the successful commercialization of digital maps online in the early years of the twenty-first century, latter services like Google Maps represent a much more comprehensive integration of mapping technologies with search engine results, as well as volunteered geographic information (VGI), digital trace data collected by Google and other online information systems, along with myriad other algorithm and analytics. In aggregate, the shift from MapQuest to Google Maps has enabled the location and tracking of users as well as the combination of maps and other spatial data with vast (and growing) troves of highly granular, and often personalized and identifiable, information. This broad-based collection of location information, and its combination with other data, in turn allows for new forms of valuation of such data in a digital environment, since it brings digital information flows into ever more explicit entanglement with the physical and social world. As a result of this development, new markets emerge at the core of the geoweb, where discrete services for static maps give way to vast and complex mapping environments shot through with digital advertising, data collection, and integration to myriad other online services. Thus, MapQuest and Google Maps are emblematic examples of two contrasting iterations of the geospatial data economy facilitated by the internet at different points in the growth and development of digital capitalism. As such, each represents an identifiable configuration of technical and economic logics, the transition between which has widespread implications. Through the analysis of the various products and services associated with these firms, and their underlying logics, the following sections document how digital maps have become central to the construction and maintenance of today's digital capitalism.

MAPS ON THE WEB AND THE LOGIC OF DIGITAL CAPITALISM

In a time of GPS-enabled smartphones, interactive car navigation systems, crowdsourced mapping platforms, and location-aware applications, accessing digital maps on the internet seems like a routine, almost unremarkable, activity. This was not always the case. In fact, the very ubiquity of digital mapping is a testament not only to its enormous success as a consumer technology, but also to the dramatic transformations it has undergone over the past two decades. After all, it was not so long ago that those who wanted to make use of online digital maps for navigation purposes had to print them on paper sheets to produce makeshift turn-by-turn roadmaps customized for each individual trip.

The roots of online digital mapping are deep and sprawling, reaching back before the internet itself or the advent of digital technologies. Yet, before examining them, it is worth spending some time reflecting on how digital maps have become so thoroughly embedded in our daily activities. At the time of writing, billions of people all over the world make use of platforms like Google Maps, Apple Maps, Bing Maps, Baidu Maps, or OpenStreetMap on smartphones, tablets, personal computers, and even cars. Indeed, today much of our daily engagements with maps, our exposure to geographic information, and our navigation through physical spaces (be it vehicular or self-propelled) is mediated by some type of digital mapping platform. Their interactivity and real-time responsiveness make digital mapping platforms practical tools for finding locations in space as well as identifying and following our preferred routes while we move through it. Yet the workings and purposes of digital mapping platforms often exceed their informational and navigational capabilities. It is by recognizing *what else* digital mapping platforms do in addition to mapping, and how they do it, that we can understand how they are different from other forms of geographic information, while also explaining why such differences matter for the integration of these products with the broader dynamics of digital capitalism.

An important feature that sets online digital mapping platforms (like Google Maps) apart from paper maps, atlases, and even from other computer-generated maps, such as those traditionally produced through GIS,[1] is that they exist in networked informational environments, upon which they are entirely dependent. While early examples of online maps like MapQuest exploited this capability only in limited ways (by featuring advertisements), newer platforms like Google Maps are comprehensively interlinked, at all levels, with ever-expanding troves of digital data circulating on the internet. This sprawling interconnection includes, for instance, the search process through which users input data into maps, the movements in physical space as they use said maps to navigate, the information that appears on the maps themselves, and even how this information changes depending on the geographic location from which the platform is accessed. These profound interconnections constitute one of the core mechanisms by which digital mapping platforms are simultaneously subsumed into digital capitalism, while also continuously reshaping it, sparking new interactions, transactions, and interconnections between digital and non-digital spaces in the process. A salient consequence of linking digital mapping platforms with other data sources circulating through the internet is affording users the ability to carry out spatial queries and find not only where locations are *in* space but, crucially, multiple kinds of information *about* those locations, which can even be provided in real time as users (physically, as well as virtually) navigate *through* them.

Thus, a mundane activity such as finding the nearest Burger King, museum, post office, or pawn shop can readily become a complex spatial inquiry involving multiple datasets (addresses and locations, user reviews, maps, real-time traffic), sophisticated spatial operations (routing algorithms, overlay of multiple qualitative and quantitative criteria), and heuristic judgment (evaluating the relative quality of the information, or contrasting the search results with personal experience or local knowledge). Leveraging these capabilities afforded by digital mapping platforms and their interconnection with other datasets and applications, users can find answers to more complex and nuanced questions that allow them to go beyond relatively simple facts about distance and location. For the burger-seeking users, this process might lead to a menu of options including the best- (or worst-) reviewed Burger King locations within a certain radius, fast-food establishments similar to Burger King, restaurants that offer some similar dishes but with an entirely different approach (say, vegan alternatives), or food-delivery services to have their burger brought to them, making the location and transportation part of the search redundant for the user—if not for the delivery worker, as discussed in chapter 1. While the number, content, and variety of these results depends on the location of the query,[2] the information retrieved, aggregated, sorted, displayed, and made accessible to the user is an artifact of the search and mapping platform itself. Due to these platform-dependent variations, the very same query (with identical location, device, and even user) made using Bing Maps may yield different results in Google Maps or any other service. In other words, the spaces produced and reproduced through the process of user interaction with digital mapping platforms are contingent on myriad factors, and continuously changing in ways that respond simultaneously to informational inputs and feedback dynamics between users, platforms, and the physical and social environment.

Of course, spatial queries can be conducted without the aid of platform mapping services, even if both the process and its outcomes might be of an entirely different kind. Indeed, people have been making complex decisions about where to get a burger for as long as burgers have existed.[3] The difference, however, is that before online maps and search engines those decisions would have been made with more limited sources of information, which were not necessarily updated in real time, and which often had to be verified by direct experience or with the aid of other technologies, like the telephone. Qualitatively, however, it is not always clear that more information is necessarily better, since different forms of organizing and understanding information in space can provide us with alternative ways of experiencing the world. For instance, in informational environments prior to the development of online mapping platforms, verifying information through in-person trial and error may have helped to update one's practical knowledge about specific locations and their broader geographic contexts. This fragmentary, intuitive, and iterative process is at the heart of how we can experience space as something whose multiplicity continuously unfolds before us, and which becomes richer through experience, interactions, time, and memory.

Thus, even before (and beyond) the mediation provided by digital technologies, humans have developed myriad ways of finding information and navigating space that do not necessarily rely on networked digital environments updated in real time. Furthermore, such spatial strategies are not limited to individuals, but can emerge from information transmitted between people through non-digital means. Think, for instance, of thickly socialized ways of learning about specific, often hard-to-find information, like a hole-in-the-wall restaurant, a pop-up store, a hidden landmark, or places of personal or communal significance. These practices can in turn illustrate some of the workings of nondominant epistemologies like *local knowledges*, or communicative processes like *word of mouth*, which can emerge through interactions between people, texts, the landscape, and other aspects of the physical and cultural environment in ways that may be informal, experiential, or structured according to logics that may not conform to those of online digital mapping platforms.

In sum, digital mapping platforms are no substitutes for other ways of knowing about, experiencing, or navigating space. Such platforms, however, have become embedded in the logics and operations of digital capitalism, exerting a significant influence on individuals, the spaces they navigate, and society more broadly. As discussed above, among the consequences of this spatial experience is that users of digital online mapping platforms can rapidly conduct sophisticated spatial operations to find highly specific locational information in real time, producing new—and often monetizable—flows of information in the process. Hence, another consequence of digital mapping platforms has to do not with their mapping or navigation affordances but with the data collection, processing, and feedback loops sparked by every interaction between users and said platforms. Such interactions are at the core of the dynamics of digital capitalism—and specifically those dynamics at the core of the LVM framework—since they are the source of ever more finely grained locational data that can be valued according to platform-specific logics and then marketized through their integration with other forms of data and their incorporation into myriad transaction networks.

Returning to our example above illustrating the role of digital online mapping platforms: the very same capability that enables users to make all manner of increasingly sophisticated burger-related locational decisions (e.g., where to get it; how to buy it; whether, when, where, and how to have it delivered, ready for pickup, etc.), simultaneously generates new rounds of potentially valuable data in the processes of searching, querying, and decision-making. These resulting data are known by terms such as digital traces, digital trails, digital shadows, or "datafied dividuals."[4] The common thread connecting these various concepts is the notion that such data are generated by users as they navigate digital environments, and are in turn captured, repurposed, and recombined by third parties for a variety of purposes, from monetization to surveillance. Thus, in digital networked environments, the process of using data, finding information, and gaining knowledge, is itself generative of more data, which is then assembled into new informational configurations and (ostensibly) leads to the creation of new knowledge. In the digital networked environments that make up digital capitalism, this cycle of data collection and transformation is punctuated by key processes of location (where is the data?), valuation (how much is the data worth?), and marketization (how can the data be bought, sold, or infused into market transactions?) Seen in this light, it then becomes crucial to understand not only how digital maps help users understand and navigate the world, but how such platforms and applications are intertwined with broader political economic dynamics. For instance, who benefits from the data produced by users? How are data, information, and knowledge generated when a user searches for a burger online transformed into something economically valuable? What other data about individuals, locations, and the links between them, might be collected in the process, and with what purposes?

As we have come to expect with increasingly complex digital environments, the answers to these questions may vary on a case-by-case basis, depending on which service, application, or indeed location, we might take as the object of inquiry. However, regardless of the particulars of each case, it would not be unreasonable to expect such answers to depend on convoluted webs of relationships between search engine companies, social media platforms, advertising firms, location data intermediaries, digital device manufacturers, crowdsourced data sources, profiling algorithms, government regulators, intelligence agencies, and perhaps illicit actors, such as hackers and identity thieves—among many others. These complex webs of data collection, circulation, commercialization, and regulation are well captured by the term *networked ecologies of location, disclosure, and privacy*, proposed by Agnieszka Leszczynski, which illustrates how these interconnected environments are constituted by spatialized relationships, themselves shaped by negotiations, exchanges, and power dynamics.[5] It is by commanding a strategic position within these networked ecologies that the technology companies behind digital online mapping platforms (e.g., Google, Microsoft, Apple, Baidu), can "connect the dots" while tracking users' movements in physical space and linking them with their profiles, preferences, and a dizzying array of personalized digital trace data. This combination of physical and digital user attributes is extremely valuable for parties ranging from advertisers and competitors, pollsters, and researchers, as well as more shadowy actors like stalkers, hackers, and spies. A fundamental part of these networked ecologies is the location of information in space, which is the connective tissue between digital mapping platforms and other forms of big, networked data. The geolocation of data, or the identification of location through various digital networked technologies—from GPS to IP addresses—constitutes a central pillar for the spatial construction of digital capitalism because it ties potentially valuable flows of information to specific, identifiable people and places, and does so within a geographic frame of reference that contextualizes that information in real-world (and often real-time) dynamics. This connection between information and location first allows for the assignment of different forms of economic and noneconomic value (associated with factors ranging from geodemographic segments to behavioral patterns) that can run both ways: data and information can become valuable due to their association with specific locations, and some locations may in turn be valued differently due to their association with information flows. To name two examples of the latter, restaurant reviews on Yelp may affect the viability of certain businesses (particularly small, family-owned ones), and certain natural amenities may suffer from increased traffic and deterioration due to overexposure on social networks.[6]

While the processes catalyzed by geolocation take place on the back end of digital applications, for users the experience may translate into an increasingly naturalized flow between nonspatial and spatial data: typing a term in a search engine (e.g., *espresso machines*); receiving results that include locations, maps, and other spatial features (e.g., ten stores with espresso machines in stock "near you"); clicking on a result to reveal its specific address and location on a map. Other aspects of this process depend on the geolocation capabilities of the device from which the user is accessing this information: while many mobile devices and car navigation interfaces are equipped with global positioning systems (GPS) with sub-meter accuracy, most personal computers use a less exact method derived from the geolocation of internet protocol (IP) addresses, whose accuracy generally ranges from city to zip-code level.[7] Device-specific configurations of technical affordances can leverage location in multiple ways that in turn engender different spatial relations: while search engine results informed by a laptop's IP geolocation may provide users a tailored offering of products or services within their general area (say, at a neighborhood scale), the GPS capabilities used by many mobile phone applications like Waze or Google Maps can give users real-time location-aware navigation instructions with turn-by-turn notifications and traffic updates corresponding to their near-exact position along a set route in physical space.

How did these digitally mediated spatial and informational experiences come into being? What technologies had to be developed for us to be able to search a universe of information and select a location to receive real-time navigation instructions? Beyond the convenience factor or the technological innovations that make it possible, it is worth reflecting on how the spatialization of digital information can shed light on the big picture of digital capitalism, a technological and political economic system that has come to exert great influence in the daily lives of billions of people in the early twenty-first century. To this end, we can highlight two key functions of maps (in this case specifically online digital maps), which are reference and navigation, and trace their evolving and expanding economic dimensions. Examining the evolution of reference and navigation in the context of specific online mapping platforms provides a throughline to illustrate how digital capitalism has been constituted as a thoroughly geographic system—even as it is often portrayed as the accumulation of placeless "flows" of information and identified with an untethered digital "cloud." In the following paragraphs I trace some of the antecedents of today's online digital maps and reflect on how reference and navigation have changed as these platforms have developed, asking what this can tell us not only about digital cartography, but more broadly about digital capitalism and its spatial architecture.

MAPQUEST THERE AND BACK AGAIN

Debuting in 1996, MapQuest was a forerunner and early leader in online mapping. The company's own history and later development constitute a microcosm of the evolving role of online mapping in the digital economy and illustrate the transformations of the key cartographic functions of reference and navigation. While MapQuest is alive and online as of 2022, its place atop the internet cartographic hierarchy has long been taken by Google Maps. Furthermore, MapQuest is today an entirely different entity in form, substance, and orientation than the firm that pioneered online maps more than two decades ago. In fact, the origins and reinventions of MapQuest are themselves evidence of the

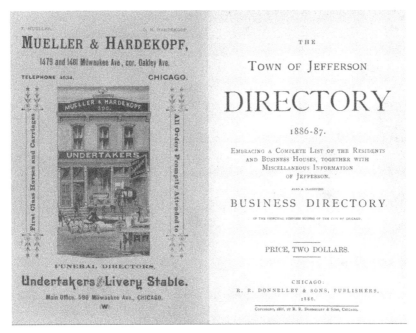

FIGURE 3. Town of Jefferson, Illinois, 1886–87 Business Directory.
Source: R.R. Donnelley & Sons, Publishers.

cross-fertilization connecting the growth and evolution of the geospatial industry, the explosion of information technology, and the many shifts, booms, and busts that have shaped the internet since the 1990s. It is thus worthwhile to dive into MapQuest's story as a way to understand where online mapping came from, how it arrived at its current configuration, and why all this is important for the construction of digital capitalism.

Unlike many of the leading digital technology companies past and present, what would become MapQuest, an early leader in internet mapping, did not have its beginnings within the internet, software, or even electronics industries. MapQuest, by contrast, began as a spinoff from a company deeply rooted in "old media": the Chicago-based commercial printing company R.R. Donnelley. Founded in Chicago in 1864, as R.R. Donnelley and Sons (Figure 3), this family firm would grow into a *Fortune* 500 corporation and is today an integrated communications conglomerate. During its heyday as a commercial printer, among other ventures it produced culture-defining informational products such as the Sears Catalog, *Encyclopedia Britannica*, and *Time* and *Life* magazines via its Lakeside Press.[8] In 1969, aiming to expand its production of maps and atlases, R.R. Donnelley founded its mapping division calling it Cartographic Services.[9] Almost three decades later, this division would come to form its own company, known as MapQuest, and become an early leader in the provision of maps on the internet.[10] To understand how a spinoff from a nineteenth-century printer came to dominate internet mapping, it is helpful to draw the connections that placed Cartographic Services within the network of innovations emerging out of mid-twentieth-century American geography and to place these connections in a broader political economy characterized by transformations in the logics and configurations of capitalism itself at the end of the twentieth and beginning of the twenty-first centuries.

An important innovation behind the development of MapQuest was the investment made by R.R. Donnelley in building a geographical database. While undertaken well before internet mapping took off, this decision would prove transformative for Cartographic Services as a company, and later MapQuest, in part because it mirrored the rising value of data as the lifeblood of the digital economy. This database was the result of a contract by *Reader's Digest* who, in 1991, hired R.R. Donnelley to create a road atlas, which was later published as *Travel Guide USA*. The process of creating this road atlas required building an original geographic database, which cost R.R. Donnelley the hefty sum of $3 million.[11] This process suggests the important role played by the assembly of databases in moving beyond purely cartographic endeavors and toward what is broadly referred to as geographic information systems, a technology, and more broadly an industry that has grown dramatically with the advent of digital technologies. Shortly after undertaking the *Reader's Digest* atlas project, and collecting the data to produce it, R.R. Donnelley's Cartographic Services division spun off into its own company, first called GeoSystems Global Corporation (GGC). This new venture took another important step toward the development of internet maps when GGC was hired by the American Automobile Association to automate the latter's routing system, called TripTik. As a result, GGC enhanced its existing database by adding routable mapping code.[12] While today such products may seem commonplace, they were in fact crucial inputs for what eventually became GGC's internet mapping service and flagship product: MapQuest.com, which launched in February of 1996.

As a first mover in the online digital mapping market, MapQuest.com captured the vast majority of cartographically inclined internet users. In fact, the service would acquire such popularity and name recognition that GGC capitalized on it, opting to change its name to MapQuest.

The newly renamed company went public in 1999 and soon after was acquired by then internet giant America Online for the sum of $1.1 billion.[13] The case of MapQuest illustrates why building a database of names, places, locations, routes, and references, became a key asset in the digital economy—even when some of these databases were created long before the commercial potential of digital technologies was widely recognized. Furthermore, the rising value of databases (particularly in contrast to other informational products, like images) also signals a profound shift in the media ecosystem and in the regulatory framework shaping the commercialization of the internet and the creation of the digital economy in the United States.

Highlighting the differential valuation across informational categories is necessary to understand the circulation and transformation of maps, and geospatial information more broadly, as they have become an increasingly central part of digital capitalism. To do this, first it must be noted that maps are not simply graphical representations of the world. While the most immediate aspect of maps may be the imagery, this conceals much of the informational inputs and processes behind it, as well as the epistemic, physical, and institutional work required to generate those inputs. Maps, therefore, contain multitudes, since they are images as well as databases, instruments as well as narratives. As critical cartographer J. B. Harley argued more than three decades ago, maps are texts just as much as they are artifacts for encoding, representing, and navigating the world. By virtue of presenting specific points of view on the world, and particular ways of framing and representing it, cartographic texts both embody and exert different forms of power over people, places, and things.[14]

Factors such as the cartographer's own perspective, the needs of clients, the requirements of concrete applications, the influence of national ideology, or the constraints of institutional rules may all affect what maps look like and what they do. Beyond this, the particular expressions of maps and their capabilities are also closely tied to the specific technological medium used to create and display them. This means that even if a map may appear identical printed on paper and displayed on a computer screen, each technological mediation shapes not only how we experience it, but also how it is constituted as an object of economic value. For instance, a map printed on a book page is well suited to foreground the graphical aspects of cartography. Yet, by the same token, the very same map may not be able to show the inputs, processes, and decisions of data collection, refinement, analysis, and visualization that precede and produce the graphical output. Digital maps, on the other hand, may open some of these aspects to the user while obscuring others. Interactive functions may, for instance, allow for real-time navigation across locations on the map, as well as zooming in and out at various scales. Similarly, the attribute data tables often linked to these maps (and which are standard part of geographic information systems) may be translated to point-and-click interfaces that produce popup balloons showing specific data points upon user request. However, these newfound digitally enabled functions do not necessarily entail more transparency regarding the process of data collection and wrangling, the relationships and data flows between the map and other digital platforms, or the (always partial and situated) perspectives embedded in graphical and technical choices such as projection, color schemes, and layout design. Thus, depending on the means and technologies we use to access and interact with maps, different aspects of them become available to us as users. Another consequence of this distinction is that, from a legal or regulatory perspective, maps may be considered entirely different objects depending on the technologies used to produce, distribute, and consume them. Consequently, printed maps, geographic information systems, and online digital mapping platforms may all be governed by different intellectual property regimes and other frameworks that guide their use, ownership, commercialization, and other potential applications.[15] The various legal regimes governing different informational products in the digital environment are a crucial component in the construction of digital capitalism, since they structure how such products can be commodified (if at all), as well as how different kinds of rights and rewards can be assigned to different actors in the digital ecosystem. Crucially, these legal regimes are also tied to particular jurisdictions, from supranational (the EU), to national (specific countries), or subnational (states within the United States), which further lends a spatial dimension to the conditions of market formation in digital capitalism.

The legal dimensions of digital capitalism are in continuous interplay with the technical configurations of the goods and services that circulate through it. In the case of maps as they shift from paper based to digital, examining how their nature, characteristics, and possibilities are linked to different technological mediations can help us understand the rise of a new paradigm such as online digital mapping platforms. This can in turn illuminate how the online mapping paradigm has become deeply intertwined with the workings of digital capitalism more broadly. As discussed by Michael Peterson, MapQuest signified the culmination of many technological innovations that made possible the distribution of interactive maps via the internet, "liberating" users from "single-scale depictions on paper," and providing individually customized turn-by-turn directions. These innovations, such as the World Wide Web and the Common Gateway Interface (CGI), which allowed interactivity with servers, help explain the rapid ascent of MapQuest as the leader in online mapping, a position that it would hold between 1996 and 2009, when it was dethroned by Google Maps.[16] Beyond the key technological innovations that made it possible, the structural features that shaped MapQuest—and later its successor, Google Maps—as a user-facing digital service should be understood as part of an emerging ecosystem increasingly characterized by commercial dynamics spreading throughout the internet, a factor that would lay the groundwork for the vertiginous expansion of digital capitalism in the first decades of the twenty-first century.

While the services provided by MapQuest itself were free of charge, the website was suffused by commercial dynamics through the presence of advertising—an arrangement that continues to be common throughout the internet, though in varying configurations. From its earliest inception, throughout its period of greatest popularity, and even in its current iteration, MapQuest's landing page has prominently featured advertisements for everything from airlines to insurance companies. While users became habituated to this ad-driven layout, feeding MapQuests's heavy web traffic for nearly a decade, eventually they began to migrate to Google Maps' cleaner interface, which in the early 2000s stood in contrast to the oversaturated aesthetic of banner-ridden websites that was the norm at the time. Like Google's search engine landing page, which featured a search box and little else, the company's service featured only a large map on the screen, which users could zoom in and out of at will—without advertised content competing for screen real estate. This visual contrast between MapQuest and Google Maps is not merely a design question but suggests a more profound shift in the political economy of the internet, since each site represents the embodiment of a particular business model, and therefore, a different articulation of the dynamics of location, valuation, and marketization of geospatial information on the internet.

To be clear, despite their significant differences in form and content, both MapQuest and Google Maps (along with related Google services) were thoroughly reliant on the business of advertising, albeit in ways that embodied specific means of intersecting with processes of location, valuation, and marketization. As will be evident throughout the rest of this and subsequent chapters, in each of the many transformations undergone by the internet over the past three decades, advertising has been a consistent throughline in the development of the commercial internet, and a central pillar of digital capitalism. As Matthew Crain has described in his history of this industry, online advertising is "the big winner" in the race to commercialize the internet, due to its "virtually unrestrained commercialization of consumer surveillance."[17]

While MapQuest was a company that drew revenue from direct advertising on its front page, Google Maps, and Google more generally, concealed this dependency from the users. In fact, by subsuming many of the functions of direct advertising into its search engine and results, Google simultaneously potentiated ad revenue while making advertisements much less visible to end users. An important driver of this shift is the fact that Google as a search engine, as a firm, and later a technological conglomerate (through its parent company Alphabet) is built on the foundation of collecting, sorting, and organizing data. This in turn allows Google products to sell astonishing volumes of targeted, segmented, and finely grained advertising. In this business model the advertising layer does not need to be visible to the end users. Rather, it is the user's own search terms, website interactions, and map navigation that provide Google services with the necessary inputs to tailor advertisements for specific profiles—a more effective (and valuable) model than one-size-fits-all advertising. This highly valuable user-generated information is then used to enrich troves of data that can then be fed into algorithms that underlie Google's advertisement business.

While data collection and advertisement customization are at the core of Google's search engine and other products, they also have an important role for Google Maps, and therefore the broader geospatial experience on the internet. More specifically, the placement and functions of advertising in MapQuest versus Google Maps point toward a change in how digital maps are constructed and navigated by the users, and how those users interact with places in physical space. While MapQuest featured "zoomable" maps and customizable step-by-step directions, its interactivity was rather limited, and each individual map was static. This meant that while users could select between ten different zoom levels, they had to do so by clicking each one deliberately. Further, while MapQuest's turn-by-turn directions functionality was very innovative for the time, it relied on breaking down each journey into separate steps, which users had to print out to use for real-time navigation. This contrasts with the interface normalized by Google Maps in later years, where users could pan throughout the map and zoom in a continuous fashion using their computer's keyboard, trackpad, or mouse. This interface became even more interactive with the widespread adoption of smartphones and tablets equipped with touch screens, which allowed users to use their fingers to search, zoom, and navigate the maps. Initially the divergence in user experience between Map-Quest and Google Maps emerged out of a key innovation in rendering maps online as multiple small tiles, rather than single images.[18] This tile-based model, known as multi-scale pannable (MSP) would later become the norm across interactive maps on the internet. Google Maps gained an advantage through this model due to its various technical advantages, which included faster loading on screen and the ability to adapt to different user queries and settings on the fly, including real-time panning in any direction.[19] By contrast, MapQuest would have to load full maps for every zoom level, which slowed down the navigation while presenting the user with pre-rendered maps that may have approximated, but not exactly satisfied, specific locational queries.

In many industries having a first-mover advantage in establishing and capturing a market can be an important factor for long-term dominance. However, in the dynamic digital economy of the early 2000s this was not enough for MapQuest, who was unable to adapt to a rapidly changing technological and economic landscape despite its initial dominance. As a newcomer that only launched in 2005, Google Maps combined new forms of data collection, personalized advertising, and improved interactivity in a way that gave this mapping platform a competitive advantage while drastically altering the digital online mapping landscape. Emblematic of this shift is the release of Google Maps' smartphone application in 2007, very soon after smartphones became available.[20] The innovations of Google Maps, together with the locational and interactive capabilities of smartphones gave rise to a new paradigm in mobile mapping and navigation. Specifically, Google Maps took advantage of GPS navigation included in smartphones to provide a real-time, interactive navigational experience where users could follow directions as they moved through physical space.

Attending to the relationships between interface design, user experience, and the changing affordances of devices embodied respectively by MapQuest and Google Maps sheds light on multiple and shifting configurations of hardware, software, business practices, use patterns and other factors at the core of digital capitalism. While MapQuest provided some interactivity and represented an important innovation in the provision of customized content responsive to user inputs (such as queries for directions), nevertheless, this service was still anchored in a paradigm of the internet that was fundamentally based on the storage and transmission of static content. In hindsight, the reign of MapQuest atop the incipient online digital mapping space seems overdetermined by the duration of the static internet. The conditions that enabled MapQuest to thrive would soon be eclipsed by what would later come to be popularly known as Web 2.0. As discussed earlier in reference to volunteered geographic information, the notion of Web 2.0 as a particular "stage" of the World Wide Web is often characterized by the enlarged role of users as content producers and distributors, and the proliferation of interactive, and later socially networked platforms and applications.

Paradoxically, the more MapQuest retained its market share in internet maps, the more the internet changed in a direction that was fundamentally incompatible with this company's approach. Thus, as the first decade of the millennium came to a close, key developments such as the rise of VGI, GPS-enabled smartphones, increasingly sophisticated interactive interfaces, highly granular user-data collection, and social networking converged to dramatically transform the digital landscape in a way that was much better suited for the strengths and orientation of Google Maps, which would soon become MapQuest's successor as the leading online mapping service.

TO ORGANIZE THE WORLD'S INFORMATION—AND MAP IT

The genesis of Google as a Silicon Valley startup, incorporated in 1998 to commercialize the research of Stanford Computer Science graduate students Sergey Brin and Larry Page is now well known, both in the Valley's lore and in broader popular culture. Along with the oft-retold stories of the two Steves (Jobs and Wozniak) forming Apple Computer Company in a home garage in Los Altos, California, or Mark Zuckerberg starting (The) Facebook from his Harvard dorm room, the Google origin story has become a foundational myth of today's digital economy. Google's role in shaping this economy cannot be overstated, nor can its impact in transforming the digital online mapping landscape. While there are many reasons for the sprawling influence of Google, the PageRank is central among them. It was this specific algorithm, which emerged out of Brin and Page's graduate research,[21] that gave the nascent company the tools to pursue its ambitious mission "to organize the world's information and make it universally accessible and useful."[22] While it was the cornerstone of Google's flagship Search product, this algorithm also signified a watershed moment in how information is collected, assembled into products, and consumed, which, since it includes vast amounts of spatial or geographic information, had transformative ripple effects for the online mapping economy.

PageRank was groundbreaking because it brought legibility, order, and hierarchy to internet searching. At the time, though there were many existing browsers, search results often did not match the queries made by the users. Across the board, search engines like Hotbot, Excite, Altavista, Lycos, Ask Jeeves, and Webcrawler—and even industry-leading Yahoo!—struggled with ways of organizing search results in ways that proved to be intuitive for users. Part of this was due to the cluttered appearance of advertisements in the search results, as well as the metrics selected to structure those results (such as word frequency, for instance). By contrast, PageRank used a network methodology to organize the websites reported in search results. Specifically, PageRank focused on the links that pointed to each website, which it then ranked by number and quality. Using these "backlinks," the algorithm inferred the importance of each website, with those that had the most rising to the top of the search results. Armed with this innovation, Google rapidly started gaining traction in the search engine market, opening the door for its growth as one of the largest technology companies in the world.

Soon the company set out to expand its collection and organization of data to new domains beyond search engines, proving the economic viability of a business model premised on collecting, sorting, and monetizing information. As Google branched out into domains like email, social networking, browsers, operating systems, social networking, and other services, the company then set its sights on the power of maps as key assets in a fledgling digital empire.[23]

While some of Google's expansion was born out of in-house innovations, much of it was the result of acquisition. Early on in its history, in 2001, Google acquired Usenet discussion services from Deja, which later were used to develop its early foray into social networking: Google Groups. As the company moved to buy blogging services (Genius Labs and Pyra Labs, in 2003), email customer support technologies (Neotonic Software, also in 2003), and digital photo management (Picasa, in 2004), it would soon make a string of purchases that not only allowed it to have a presence in the internet mapping sphere, but would come to profoundly influence how maps and geographic information are integrated into the digital economy more broadly.

Between 2004 and 2006, Google brought together via acquisition a series of technologies that would later be combined to form the core of its mapping and geospatial operations (which today include Google Maps, Google Earth, Google Street View, and Google Earth Engine). In 2004, Google bought Keyhole Corp, a digital company initially funded by InQTel, the CIA's venture capital arm, and which would later form the basis of the 3D satellite imagery visualization platform Google Earth.[24] That same year, Google acquired Australian mapping startup Where 2 Technologies, which served as the basis to develop Google Maps. To round out the geospatial acquisitions for 2004, in December of that year they purchased ZipDash, which provided maps and traffic for mobile devices. With these building blocks, Google began putting together a comprehensive suite of applications that extended its mission from organizing information from the web to organizing different types of geographic information about the world itself. In other words, Google took an explicitly spatial turn toward organizing and mapping the surface of the Earth from a variety of perspectives—from satellite imagery and aerial photography to vector maps, street-level photography, and turn-by-turn directions. This initial round of purchases was later complemented by the acquisition of Endoxon, another mapping software company, in 2006.[25]

By bringing together the technologies from the purchases discussed above, Google's expansion into the geospatial field changed the way maps circulated on the internet and how they were experienced by the users. This had broad ramifications by making digital maps and geographic information (including location data) more accessible, more interactive and, crucially, more interconnected with other online services, all of which strengthened the links between information circulating on the internet, and the places, spaces, and locations where it could be valued in economic (and other) terms. By strengthening this linkage and allowing new forms of valuation of spatialized data, Google positioned itself at the center of new digital markets that, fueled by search and targeted advertising, came to depend significantly on the geographic dimensions of digital information. (See Map 1.)

The geographic shift in Google's approach, and its most salient incarnation can be found in Google Maps, its principal mapping product. Launched in 2005, Google Maps represented a paradigm shift for internet maps. While some of the technical innovations discussed above set this product apart from leading competitors, especially MapQuest, Google Maps signaled a more profound transition in the way maps were embedded with the growing (and changing) universe of information

MAP 1. Google Maps circa 2023. *Author:* Google

that circulated on the internet. One way to illustrate the significance of this change is to consider the contents of the maps themselves. While MapQuest had respectable coverage in North America (and struggled in many other parts of the world), the primary function of this product was to serve as a spatial reference and provide route directions to users. Google Maps, on the other hand, approached mapping differently in part because it belonged to an informational ecosystem that was built around Google's stated aspirations to organize (and monetize) the entire world's information.

Accordingly, while users' initial purpose in consulting Google Maps may have been to look for geographic references or specific route directions, unlike existing products, these functions were interconnected with those provided by Google's growing "suite" of products and services—principally its search engine. While this integration between search, social networking, advertising, email, calendar, and other functions was a gradual (and is still an ongoing) process, the difference in orientation set Google Maps apart from its competitors, endowing it with greater informational density as well as cross-fertilization between informational categories that static internet maps could not provide. It was through the combination of these qualities that the contents of Google Maps (such as sites, names, locations, landmarks, addresses, and, increasingly, user-generated content like reviews) could serve simultaneously as a reference point and a portal to information collected and organized by Google via its other products, chiefly the search engine.

This integration between Google Maps and Google's other services, and information on the internet opened new horizons for the online advertisement industry. Among others, ad content could be linked to search results about particular locations, embedded in the maps, and, perhaps more importantly, linked to user location and profiles. These possibilities offer a fundamentally different approach to the informational architecture of maps deployed by Google's predecessor, starkly contrasting with MapQuest's relatively static and self-contained qualities. Integrating maps with search and other services provided by Google not only provided a more popular and practical user interface, but it also linked digital information with the social and physical landscape at an unprecedented scale. This vast linkage is the core of the spatial construction of digital capitalism, since it enables (often economically generative) interplays between the characteristics and activities of concrete sites, locations, and places, with the information, user decisions, behaviors, and traffic patterns that circulate through digital information networks. Furthermore, this interplay works both ways. On the one hand, Google Maps' integration with other Google products helped imbue places, spaces, and locations with all manner of digital referents collected from online sources—such as websites, news mentions, audiovisual media, official government information, user-generated content, and so on. On the other hand, it attached geographic counterparts to online content—such as search words, websites, and comments. Importantly, by tightening the link between digital information flows and geographic space at a vast scale and in a dynamic way, Google Maps opened a more direct portal for the valuation and commercialization of the physical locations contained within this platform, as well as the information that referred to those locations in some way. In other words, by organizing location, providing an online mapping platform, and connecting this platform with other online services in a dynamic and systematic way, Google Maps catalyzed the commodification of vast volumes of geospatial information and the creation of markets where this information circulated as part of economic exchanges. This in turn affected what appeared on the map and how users used it to navigate both in virtual and physical environments. In this way, digital online maps (particularly those created and hosted by ad-driven internet giants like Google) have become a terrain where the information attached to real-world locations can generate different forms of economic value.

From a user perspective Google Maps represented a departure from the online services common in the internet of the mid-2000s. While

MapQuest provided zoomable, but largely static maps, this experience was fenced off from other functions of the internet, aside from the presence of relatively clunky advertisements. By contrast, when a user searched a term via Google, they might be directed to Google Maps, which could give them directions to reach a physical location related to their search. Alternatively, businesses and other websites can link directly to Google Maps using the application programming interface (API), which increases their visibility. Furthermore, the information on the map itself is not static, and often features specific labels, landmarks, and highlights that can directly or indirectly boost traffic or interest through visibility and exposure.

As Google Maps has developed increasingly sophisticated tools and integration with other forms of data and platforms, it has become an important part of digital marketplaces both on the internet and beyond. For instance, the development of Google Street View has catalyzed not only online real estate services like Zillow, but has paved the way for Google's move toward developing technologies that can benefit from vast amounts of street-level imagery and other forms of geospatial data, such as Autonomous Vehicles.[26] I take up some of these developments and the link between geospatial data and mobility in chapter 4. Before that, in the following chapter I interrogate the multiple facets of the idea of location and how they are linked to geolocation tools and the allocation of different forms of value, a process at the core of digital capitalism.

CHAPTER 3

Location, Geolocation, Allocation

This chapter addresses the rise of geolocation, geotagging, and other practices through which information flows (web searches, social media content, individual computer profiles, personalized user behavior) become linked to concrete geographic locations such as coordinates, identifiable places, or street addresses. The chapter first establishes the importance and multidimensionality of location along with a framework to understand it as a way of simultaneously finding things in the world and siting them. Having established this framework, the chapter then explores the incorporation of geographic location into digital flows through techniques of geolocation. Having contextualized the rise of geolocation technologies, the chapter then proceeds to explore their larger political economic implications for the development of digital capitalism. Crucial among these implications is the ubiquity of digital goods and services increasingly tailored by geographic and demographic factors, as demonstrated through applications whose offerings change depending on the location of the user, like recommendation engines (e.g., Amazon), review websites (e.g., Yelp), geographically determined video libraries (e.g., Netflix), individually tailored advertising and content provision, and navigation services.

LOCATION

Location is everywhere, and everything is tied to location. It is indeed almost impossible to find any aspect of (human and nonhuman) life unrelated to, or unaffected by, location. *Where* we are located is closely tied to *what* we can do at any given time. Over the course of our lives, *where* we are becomes integral in shaping *who* we become, as is the case with the influence exerted on our lives by the neighborhood we grew up in, the school we went to, or the job we have. Conversely, *who* we are is often instrumental in defining the range of possibilities of *where* we can be. The association between location and other aspects of our lives is not absolute or deterministic, but rather mediated by myriad contextual factors such as the type of society we live in and its structuring institutions, the physical environment, the degree of spatial and social mobility, and axes of difference and power such as race, class, gender, disability, and immigration status.

Location is not destiny, but it is a point of departure that exerts influence on the paths we chart both in time and space. At the same time, location can help make sense of where we've been in the past. Where we are located, for any length of time, can have small or large impacts on our lives and trajectories. On a small scale, it is location that determines whether we can safely cross the street, seek shelter from the elements, or step out of bounds in a basketball game. And while location can be identified by a pair of coordinates on a map, it is not reducible to them. Location, then, is relational because it cannot be fully understood as an isolated fact, but only in relation, interconnection, and interdependence to others—other locations, other people, places, actors, and always embedded within a broader frame of reference. In fact, there are many ways to make sense of location depending on our frame of reference, whether that is in the process of interpreting a scan of the human body to find an injury, the exploration of the deep sea to map its terrain, or a space telescope capturing the depths of the universe from its position around Earth's orbit.

When we trace its myriad implications, it becomes clear that while location can be many things, it is inherently (and intensely) political. This is because the visible and invisible webs of power relations shaping who gets what, when, and how are always embedded in concrete locations and spatial contexts, which in turn have an active role in weaving those webs.[1] The ability to access, occupy, and control space (or to deny others from doing so), to ascertain someone else's movements, or to move undetected, to gain knowledge about a place, or to prevent circulation of that knowledge—all involve different power geometries that are irremediably connected to location. In the context of territorial expressions of power, such as immigration, foreign relations, policing, trade, and war, location is a key nexus that can determine the course of events, often with significant implications. It is by identifying location at a given moment that one can establish whether something or someone is within the borders (or air space, or territorial waters) of a bounded political area, like a country, and therefore subject to its jurisdiction. Alternatively, through location we might also know whether one is outside of said county's boundaries, in international space, or even in disputed territory. Furthermore, the relational and contextual aspects of location can then help determine how power is exercised. For instance, whether an actor's presence in physical or social space is considered legitimate, or "unauthorized" by the governing authority of a particular territorial jurisdiction. Depending on where, when, and even, interpreted by whom, one can be "behind enemy lines," "within state limits," "in international waters," "in no-man's land," "in public space," or "on private land"—all of which are thoroughly political expressions of location.

It is by determining the meaning and implications of location that specific actions can be allowed or forbidden: if someone is alleged to be trespassing on private property, the owner of said property may or may not take some kind of action against the trespasser, and the range of legitimate actions is a direct consequence of the normative and legal framework that applies in that particular location. This illustrates how, even as location contains and reveals a wealth of information, its meaning cannot be fully grasped without interpreting it in context. A location's context may involve, in the first instance, its relation to a broader geographic setting—a surrounding or containing space (i.e., *where is this location*)? However, context is also made up of the rules governing the geographic setting of a location, which can establish, for instance, where we are (or are not) allowed to set foot. In this case, the consequences of setting foot on someone else's property can be radically different whether one is in the US state of Texas or almost anywhere in Sweden. This means that location is inseparable from the explicit or implicit exercise of power—often, though not exclusively, backed by the state. Yet, while formal rules, like legal frameworks, are key to establishing and legitimizing it, a location's context is influenced just as much by informal norms, values, assumptions, and cultural practices. This is how, beyond the differences in legal frameworks governing Texas or Sweden, the ramifications of a trespassing incident cannot be separated from the power asymmetries between the parties involved, and how these are intertwined with social cleavages like classism, gender discrimination, or racism. This mix of contextual factors, in different measure, interact to infuse locations with meaning and consequence; in this case informing who gets to have the presumption of guilt or innocence when in a certain location—a thoroughly political outcome, since it establishes who can (or can't) do what, where, and at what cost.

In the examples above, location is a site of contestation where various actors interact under conditions set by a governing system of rules, values, practices, and behaviors. However, even in the absence of direct or visible contestation between identifiable actors, location is intrinsically tied to its surrounding environment and interacts with it in profound ways. This relationship brings up a different kind of politics: what does something or someone's location tell us about the distribution of "goods," or "bads," in the world? As it does for millions of people around the world, particularly in poor and marginalized communities, location (of a settlement, a home, or one's own body) can mean exposure to toxic substances, polluted water, or high concentrations of atmospheric particulate matter. By contrast, different locations can mean abundance of green areas, clean water, shaded streetscapes, and breathable air. Understanding location in context, then, means also coming to terms with the fact that neither of these examples is the product of a neutral or "natural" environment, but they are both historical and geographical constructions deeply intertwined with the systems of power, domination, and social organization that produced and continue to reproduce them. Put another way, the problems of justice, environmental and otherwise, cannot be divorced from the problem of location. Therefore, how we choose to incorporate location into decisions about governance, resource allocation, and administration is fundamental to shaping what location means and what it does. For instance, while in Finland schools are "nationally funded based on the number of students," in the United States "public schools are locally funded, usually from property taxes, and rewarded based on high performance through programs such as the US Department of Education's Race to the Top grants."[2] This example illustrates how the location of a family's residence can have vastly different implications for their children's education and opportunities depending on the type of society they live in and how that society incorporates location (and the factors associated with it) into the allocation of goods, services, benefits, and harms. Yet, as consequential as location is, this example also illustrates how its meaning and consequences are not set in stone, but are a matter of policy, political will, choice, and power, as much as chance.

On a different register, location is not only political at the interpersonal and collective levels but can also be tied to our inner personal lives.

Location, that is, can also be emotional and experiential. Specific locations can trigger the vast affective repository of our lived experience, merging physical space with experience and emotion: the site of a first kiss, a fond memory, or a last goodbye are likely to carry emotions and meanings accessible in their full measure only to ourselves, but perhaps legible to others. Adopting this approach to location, the online collaborative project *Queering the Map* has built a rich repository of memories and experiences of queer life volunteered by users all over the world. These are pinned to locations on a digital map, along with written descriptions suggesting their personal significance. The project is described as "a community generated counter-mapping platform for digitally archiving LGBTQ2IA+ experience in relation to physical space.... Through mapping LGBTQ2IA+ experience in its intersectional permutations, the project works to generate affinities across difference and beyond borders—revealing the ways in which we are intimately connected."[3] The result is an interactive visualization of a layer of human experience that is often invisible, suppressed, or devalued, yet whose existence and public representation transforms locations all over the world by enriching their meaning. Locations are therefore only partially "points in space," since they are simultaneously meaningful, multilayered, interconnected, and deeply entwined with geographic context. For geographers, it is this experiential, symbolic, socialized, and meaning-making dimension that can transform locations into fully realized places.

The richness and possibility of location are not lost on those actors with the capabilities of locating things in the world and generating some kind of value from them—from cartographers to intelligence agencies, national governments, and social media companies. The act of locating can involve two separate but related undertakings. The first (finding) is to identify where things are located, which can help us explain why things are located where they are. This is in turn connected with a second sense of location (siting), which is deciding where things ought to be, and acting upon it by setting or establishing something somewhere. Geographers are concerned with both in different measure, each involving its own scholarly traditions and ways of thinking. Economic geography has for more than a century built a body of knowledge about the first sense of location by aiming to explain why human settlements, and other sites of economic activity are located where they are. These attempts gave rise to location theory in the nineteenth century, which was variously expanded on, refined, and later incorporated into the broader frameworks of spatial and regional science, which were fundamental to the discipline's quantitative revolution around the middle of the twentieth century. Explanations on why things are located where they are included models for patterns of land use, settlement, transportation, and trade.

Some of the foundational models in this body of work were developed in Germany in the nineteenth and early twentieth centuries: from von Thünen's concentric circle model to (Alfred) Weber's industrial location model, and later Central Place Theory, developed by Walter Christaller and August Lösch. These models sought to systematically account for the hierarchies, relationships, and connections linking different types of locations across a given area (often modeled on the paradigmatic featureless plain). Among others, these models sought to explain the relationship between larger cities and smaller towns as well as how those settlement patterns were influenced by the spatial distribution of factors such as natural resources, transportation networks, and later the construction of markets for commodities, goods, and services. This knowledge was not only used for descriptive purposes, but was also closely intertwined with policy and state action. The range of applications of these models is vast, covering areas from urban, regional, and industrial planning, as well as corporate strategy. While these are all expressions of power in their own right, perhaps the starkest illustration of the power of location models is in how they have been employed for destructive ends. Concretely, location models have also been fundamental for the appropriation and transformation of lands, expulsion and extermination of populations, and colonization projects that seek to shape and control the world in their own specific ways. As Trevor Barnes has extensively documented, this was the case with Walter Christaller's hexagonal model, which became instrumental for Nazi Germany's Eastward expansion:

> Christaller's task was to plan the newly Nazi-annexed territory of western Poland in conformance with his central place theory, which he set out in his doctoral thesis in the early 1930s at the University of Erlangen. For western Poland to be transformed into Christaller's central place model, most of the region's residents, primarily Jews and Slavs, were forced to leave; many of them were sent to their death. The now "empty space" of western Poland was "reterritorialized" as a German central place by importing *Volksdeutsch* immigrants.[4]

Central Place Theory is illustrative of the entanglement between location knowledge and the exercise of power as well as its most extreme consequences. Yet the same theory also stands as an example of how such patterns can be resisted and disrupted. As Barnes notes in his discussion of "a morality tale of two location theorists in Hitler's Germany," the story of Christaller's Nazi collaboration can be productively contrasted with that of August Lösch, the other leading scholar behind Central Place Theory. Unlike Christaller, "Lösch was able to assert moral agency, never joining the Nazi's, never swearing allegiance to Hitler, and being subversive when possible." In part, this was due to the institutional conditions that allowed him this possibility: "He was able to do that in part because others, in this case his 'director,' Andreas Predöhl, chose to work with the Nazis, to become one, to join the party, but which then allowed him to protect some of his dissenting staff including Lösch; that is, to allow Lösch to make his moral choice."[5] This tension surrounding one of the cornerstones of location theory and regional science should serve as a reminder that it is not only location itself that is political, but the tools and methods through which we conceptualize, determine, and act upon it are inseparable from power and politics at all levels. For these reasons it is important to remember that establishing a location (both in senses of dictating and determining it) is necessarily a political action, and therefore it is always situated in concrete historical contexts and mediated by specific institutions and other social structures.

Academic disciplines constitute another such structure mediating the production, distribution, and application of knowledge about location. Throughout the rest of the twentieth century, location theory and regional science fluctuated in terms of its prominence in geography. In part, these shifts were themselves the result of contestations over the ties between location, the tools we use to measure and determine it, and larger structures and systems of power, from the state to global capitalism. The postwar decades saw methods and models of location theory, regional science, and spatial science more generally expand in sophistication and influence, aided by the availability of computers, a demand for positivist approaches to planning, and the broader quantitative turn in the social sciences. However, from the 1970s, the crisis of capitalism itself brought with it a "Marxist turn" against these approaches in economic geography.[6] In part, this was due to the growing sense that location theory and associated quantitative, formal, and analytical approaches were incapable of explaining fundamental social processes and their outcomes, many of which were reaching a boiling point through the contestation of racial hierarchies, economic stagnation, class conflict, energy crises, and more generalized social upheaval.[7] This signaled a break in geography between those scholars in the GIS and geospatial science community who embraced the tools and methods of regional science, and many human geographers who viewed quantitative tools, methods, and approaches to location with considerable skepticism. During the 1980s and 1990s, a subset of discussions about the role of geospatial data and geographic information systems came to be known as the "GIS Wars." These (at times heated) debates sought to address the different epistemic perspectives and political dimensions of new mapping and spatial science techniques, and their tension with other paradigms in geography such as Marxist, feminist, and postmodern theoretical frameworks.[8]

More recently, however, quantitative approaches to location and spatial science have once again gained momentum in economic geography, human geography more broadly, and adjacent fields, taking advantage of dramatic expansions in the availability of big (geospatial) data, increased computing power, and sophisticated geospatial analysis tools—from GPS-enabled mobile applications to the popularity of web mapping and the increased availability of high-resolution (spatial and temporal) Earth observation remote sensing data. While such approaches have not necessarily become hegemonic in economic geography, they are nourished by a rising popularity of computational tools, data science, and other quantitative approaches across the natural and social sciences, and even the humanities. Among the key problems recently addressed by this wave of scholarship in economic geography are the relationship between patterns of location and the development of knowledge, ideas, cultural production, and technological innovation, all particularly germane to the digital economy of the twenty-first century.[9]

In the intellectual traditions of regional science, location theory, and spatial analysis, as well as geographic information science more broadly, location is a key variable that underpins systems of spatial relations, which themselves form the underlying structure of human and non-human activity as it unfolds across the Earth's surface. Location thus becomes both something to be explained (why is Los Angeles located in a decidedly environmentally inhospitable site for a great metropolis?)[10] as well as part of the explanation (Chicago's location and its vertiginous urban growth during the nineteenth century were mutually reinforcing factors that led to the city becoming the "gateway to the West")[11] in the process of making sense of why the world is the way it is. Of course, as evidenced by the examples above, it should be clear that location and economic activity are not static but continuously reshape each other. That is, while in some cases, the initial location of something (such as a human settlement) may be explained by the spatial distribution of preexisting endowments (like natural resources, climatic conditions, or proximity to waterways), the absolute and relative significance of such endowments may shift over time. New endowments may in turn result from previous developments at this particular site, which can effectively change the character of a location by transforming the space around it while also changing its relationship to other locations. Similarly, due to factors like technological innovation, energy transition, a shift in the system of production, environmental deterioration, or some other paradigm change, what were once seen as endowments (like coal fields) may become less significant, or even detrimental, at another time. This co-constitution between location, human, and nonhuman activity can help explain why the fortunes of places are not static and solely determined by some predetermined geography, but rather that they are engaged in a dynamic process where location is influenced by geography just as much as it exerts an influence upon it. For instance, a place that may have been considered economically unsuitable during one historical period may become, under different historical circumstances, a magnet for new immigration in another period. Viewing the relationship between location, geography, and history as characterized by continuous and mutually reinforcing dynamism can help us understand the ebbs and flows of places over time, from towns and cities to nation-states and empires.

Take, for instance, the reversal experienced by many European countries over the past century and a half, which went from net migration-sending societies in the nineteenth and early twentieth centuries to net receiving destinations in recent decades. According to economic historians Ran Abramitzky, Leah Platt Boustan, and Katherine Eriksson, "During the Age of Mass Migration (1850–1913), the United States maintained an open border,[12] absorbing 30 million European immigrants" in "one of the largest migration episodes in modern history."[13] In this population movement of world-historic proportions, Ireland stood out as a particularly extreme case. While emigration was a part of Ireland's history at least since its colonization, first by the English, starting in the mid-sixteenth century, and then the British Empire, this trend accelerated dramatically in the nineteenth century, especially in the aftermath of the Great Irish Famine (1846–1852).

Data from the EMIGRE Project at University College Cork show that "approximately ten million people have emigrated from the island Ireland since 1800."[14] In locational and relational terms, we can read this population movement as a story of how the decline of one place was connected in deep and complex ways to the rise of another one. Geographer Dylan Connor explains, "Emigration contributed to the decline of the Irish population from more than eight million in 1840 to less than four and a half million by 1913 and was precipitated by sluggish domestic economic development and expanding opportunity in the United States."[15] While the scale of migration reduced in the following decades, Ireland continued sending more migrants than it received well into the twentieth century. Yet a stark reversal of this historic trend has taken place since the mid-1990s, when Ireland became a net migrant receiving country in nearly all years from 1995 to 2022—except for the aftermath of the Global Financial Crisis (2009–2013)—peaking in 2006 with a net gain of 94,996 migrants.[16] The rapid economic growth in the 1990s and 2000s during the period known as the "Celtic Tiger," Ireland's European Union membership, a friendly tax regime for corporations, and other site-specific factors, along with regional and international trends in an age of accelerated globalization have combined to make Ireland (in net terms) a migrant destination, rather than a point of origin for massive emigration.

This stark reversal in Ireland's recent history illustrates how locational factors, and seemingly fixed endowments "in place" are in fact dynamic, shaped by broader historical and geographic contexts, and related to events and developments in other locations, all of which means that they can experience shifts—subtle as well as dramatic—over time. It would be foolish to try to deny the lesson contained in the emphatic epizeuxis of "location, location, location," perhaps the number one rule in real estate (for good reason). However, as I suggest above, and elaborate below, not only is location not destiny, but the meaning and value we assign to location are themselves subject to change, among other things, due to the technologies we use to determine it.

Even as location and economic activity reshape each other, together they can still generate patterns that may become entrenched over time, thus setting places on trajectories that might be difficult to change. The strength of these patterns and their enduring quality can be expressed through concepts like circular and cumulative causation, path dependency, and "lock-in," all of which characterize how conditions in a location at one time can influence the likelihood of future events, which in turn makes some outcomes more likely, and others less so. Thus, for instance, in the example mentioned above, the initial conditions that led to Los Angeles becoming the hub of the US film industry are in hindsight not particularly unique or dispositive. In fact, at the turn of the twentieth century when the film industry was initially located in New York City, there were many cities vying to attract filmmaking. Key advantages often ascribed to Los Angeles, such as warm weather for year-round filming, were shared by other competitors like Jacksonville, Florida, nicknamed the "Winter Film Capital of the World." However, once Los Angeles gathered a critical mass of studios, movie stars, and connected businesses, it also began to develop competitive advantages through innovations in filmmaking, such as the studio star system and more complex narrative forms for feature films. All these factors, along with the economic dynamism and growth of Los Angeles itself, combined to develop unique local dynamics that could not easily be replicated elsewhere, establishing a true locational advantage—one that was created, rather than given. In this way, L.A.'s path to growth was "locked in," leading to the emergence of Hollywood as the leading global hub in the entertainment industry.[17]

While the relationship between location and economic activity may fluctuate unpredictably over time and space, both are closely linked in multiple ways. Yet it is worth spending a moment to examine how this link is established. Regardless of the endowments or qualities of any given location, for them to exert any kind of influence (at least on a human level), they must somehow be known, communicated, and incorporated into decision-making. Therefore, we cannot understand the importance of location without also coming to grips with how we identify, record, document, interpret, analyze, and act upon location through data, information, knowledge, and the myriad technologies that make all this possible. Location is therefore not reducible to the position of something in space but is rather multifarious and multidimensional. Key to the construction of location is its informational dimension, by which we come to know it, communicate it, endow it with meaning and value, and act upon it in myriad ways. It is by acquiring information about a location that we can assign to it symbolic qualities, make relevant political decisions, channel (or extract) resources, and associate it with the production of economic value. For instance, over the course of millions of years, a crude oil reserve may have been quietly gathering deep underground below a particular location. Yet it is through new information about the discovery of the reserve that the significance of the location can change, with transformative economic and environmental consequences. In this case, such consequences would often depend on the ability to conduct oil surveying to develop maps and models showing where a reserve may be located. This, in turn, is not sufficient to change the fate of this hypothetical location, since what is done with the information about the oil reserve in turn depends heavily on other contextual factors, such as the governing regime and its energy policy, the relevant legal framework, popular or political sentiment surrounding fossil fuels, and negotiations about the future extraction of the resource and the distribution of potential revenues, among others.

Thus, decisions about *where* to locate things in the world (in both senses of the term, finding and siting things in space) are closely connected to *what* we know about *where*. By the same token, *what* we know about *where* is inseparable from the technologies through which we navigate and make sense of the world, as well as the technical and informational environments in which such technologies circulate. Location technologies may change over time, but they should always be understood within the broader context of a political and economic system that structures their conditions of possibility. Thus, if we want to understand the role that location plays in digital capitalism, it is necessary to bring into conversation the ideas developed throughout the preceding paragraphs with the specific techniques used to incorporate location into digital environments. The following section takes up this discussion through an exploration of a particular set of technical infrastructures and methods for locating things in the world that has become central to the functioning of the digital economy, geolocation, and discusses its implications within the framework of *location, valuation, marketization*.

GEOLOCATION

The paragraphs above explored the notion of how two senses of location (finding and siting things in space) feed back into each other in consequential ways. In this section I argue that as new technologies for determining location develop, they are accompanied by new forms of valuation, and consequently new markets—all of which in turn affect where things are located in space. As mentioned earlier, while economic geography has devoted much attention to explaining the second sense of location by asking, for instance, why cities, industries, and other forms of spatial agglomeration are sited where they are, other areas of geography have centered tools for finding things in space. In this respect, fields cartography and geographic information systems (GIS) have developed the technical, technological, representational, and epistemological means by which we learn where things are located in space, and how we organize, visualize, and interpret this information. In different ways and with varying techniques, each of these makes use of maps, geospatial information, and often a specific subset of it, which is location data, to understand the world. Traditionally the purview of cartography, maps can be understood as artifacts made by humans to make sense of the world and find their way in it. Acknowledging the multiple ways in which humans experience and understand space, the definition of a map can encompass a vast diversity of cultural artifacts and practices as well as techniques and technologies: from atlases and paper maps to murals, scrolls, engravings, navigation charts made of sticks by Marshall Islanders, or portable wooden figures in the shape of coastlines carved by Inuit peoples of Greenland.[18] Yet, across the vast universe of maps and their myriad components, a recurrent (if not universal) feature involves identifying and organizing information about particular locations in space.

While mapping has been a part of humanity for thousands of years, it has recently undergone a profound and unprecedented shift. Among the many effects of rapid expansion of digital networked technologies is the growth of the geoweb, which is itself part of a decades-long transformation of maps, mapping, and geospatial data that can be traced back to early geographic information systems in the 1960s. While today online cartographic platforms like Google Maps, OpenStreetMap, Waze, and TomTom are a seamless part of our everyday lives, these recent innovations have redefined in fundamental ways core aspects of what maps are (e.g., interactivity and networked digital distribution), what we can do with them (e.g., real-time navigation), and how they integrate with other aspects of our lives and the political economy structuring our societies (e.g., data collection and targeted advertising). Behind these specific applications lies the combination of two key factors: first, the digitization of maps, and second, the explosion and circulation of maps and other forms of geographic data on the internet—in other words, the geoweb. The changes that gave rise to the geoweb, however, are part and parcel of broader political economic realignments involving new mapping and information technologies, private corporations, user communities, and the state. As Agnieszka Leszczynski has argued, this realignment must be situated and understood "within the neoliberal political economic restructuring of the state." This is because, while mapping had long been a primarily state-centric activity, "as the state is 'rolling back' from public aspects of the cartographic project, market regimes of governance are simultaneously 'rolling out,' subsuming the mapping enterprise to the imperatives of technoscientific capitalism."[19]

Thus, to understand the dynamics of the mapping revolution and the growth of the geoweb it is necessary to tie the networked and interactive conditions enabled by the internet with the development of GIS over

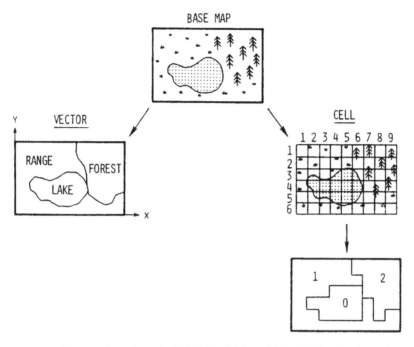

FIGURE 4. Diagram from the 1985 *MOSS User's Manual*. MOSS (Map Overlay and Statistical System) was a data analysis and geographic information system commissioned in the mid-1970s by the US Fish and Wildlife Service and developed by the Western Energy and Land Use Team. *Publisher:* US Department of the Interior, Bureau of Land Management, Division of Advanced Technology.

the past half century. GIS refers to the computer programs and packages that allow for the collection, organization, analysis, and visualization of various kinds of spatial data, including location. These packages were first developed in the 1960s, when governmental institutions and universities began to incorporate computers in their research, mapping, and resource management projects. It was in this context that key sites like the Canada Land Inventory undertaken by the government of Canada and the Harvard Laboratory for Computer Graphics and Spatial Analysis made important innovations that would give rise to myriad mapping and GIS applications, many developed by government agencies (see Figure 4 and Map 2).[20] In the decades since, these tools have underpinned the development of a new field of knowledge, called geographic information science, defined by prominent geographer Mike Goodchild as "the science behind the systems," which is driven by foundational intellectual questions, rather than technical developments.[21] Geographic

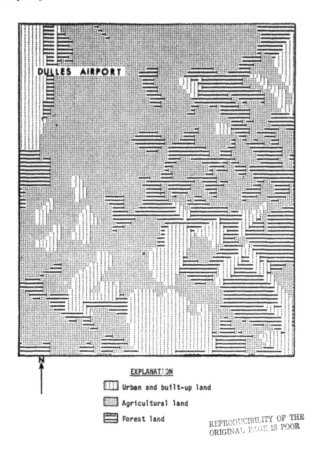

MAP 2. Central Atlantic Regional Ecological Test Site (CARETS) was a system developed jointly by NASA and USGS in the 1970s that integrated remote sensing and GIS. Image shows a CARETS computer map of land use around Dulles International Airport, Fairfax County, Viginia, 1979. *Author:* Robert H. Alexander, US Geological Survey.

information science endeavors to develop methods and theories to analyze, interpret, model, and explain geographic patterns and processes in the world with the help of geographic information systems. It combines elements from geography, cartography, computer science, statistics, and remote sensing, among others.

While a close account of the development of GIScience as a field of knowledge is beyond the scope of this chapter, it will suffice to note two key points. First, that it was spurred by the digitization of geographic information, which went hand in hand with the evolution of computer programs to process it. Second, that location information was at the core of these developments, even if this data category has only become much more salient in recent years due to the big data revolution unleashed by the internet. Both points can help explain the continuities and changes that characterize the current digital economy, the close connection between new informational tools and new epistemic and organizational developments, and the shifting role of location data in comparison with the previous decades of development of GIS and GIScience (roughly from the 1960s to the early 2000s).

Geographic (often also called spatial, or geospatial) information, encompasses a broad range of formats, data models, datasets, and informational categories; from satellite imagery, vector geometry and topographic data to digital elevation models, GPS trace data, and even geotagged social media data. The common element among all these seemingly disparate categories is that they contain some reference to, or representation of space, which can then be used to answer questions and generate knowledge about (among other things) geographic patterns and processes. However, regardless of the specific data in question, asking geographic questions and answering them within GIS is generally—though not always—enabled at a fundamental level by a correspondence between the data and identifiable, concrete locations in the world.[22] This correspondence in turn is possible because geographic information exists within some frame of reference that systematically organizes and partitions the surface of the Earth, such as the coordinate reference system (CRS).[23]

Therefore, one of the key features of GIS and GIScience is the (implicit or explicit) process of *locating* digitized data by placing them within a frame of reference on the surface of the Earth. Traditionally this process goes by the name of *georeferencing*, which, according to the United States Geological Survey "means that the internal coordinate system of a digital map or aerial photo can be related to a ground system of geographic coordinates. A georeferenced digital map or image has been tied to a known Earth coordinate system, so users can determine where every point on the map or aerial photo is located on the Earth's surface."[24] In recent times, with the proliferation of new forms of geographic information, especially that which circulates through digital networks, a new form of georeferencing has become increasingly prominent: geolocation.

Geolocation can be thought of as a way of georeferencing a kind of information whose geography is too often elided: flows of digital information on the internet. Thus, the problem of georeferencing, which has been central to the development of GIS and GIScience, and which had been largely resolved for many kinds of traditional geographic data, acquired new salience as the spatiality of information on the internet became the focus of inquiry. How this happened and why it matters to the digital economy are the focus of the following subsection.

Geolocating the Invisible

In the context of digital environments, the idea of location poses a persistent question, a nagging uncertainty: can digital flows of information be located in physical space? How can we make sense of the fact that those digital flows, which are not visible to the naked eye, may or may not exist in the same place and time as the people producing or consuming them? Asking these questions brings to light how, even as it is so often overlooked, the problem of location is in fact instrumental for every dimension of the digital economy: from underpinning its underlying informational infrastructure to bringing digital information flows into concrete social and political contexts, and shaping the imaginaries through which we perceive, interact with, and even regulate this economy.[25]

One of the earliest conceptualizations of what would become the internet, and which reveals the deep embeddedness of digital environments in geographic territory and geopolitical spatial imaginaries, was articulated by RAND Corporation engineer Paul Baran in 1964 as a "distributed communications network." In contrast to centralized networks (where all links had to pass through a node that acted as a central hub), or even decentralized networks (with many hubs), Baran argued that *distributed* networks did not have any hubs, making all the nodes roughly equal in importance and centrality. For him, this network structure demonstrated "payoff in terms of survivability for a distributed configuration in the cases of enemy attacks directed against nodes, links, or combinations of nodes and links."[26] Along with other key ideas, such as packet switching and methods to analyze data packets, Baran's conceptualization served as the basis to develop the computer network called ARPANET, which was declared operational in 1971. Unlike existing telecommunications like the telephone or the telegraph, this computer network was endowed with the capacity to reroute information should

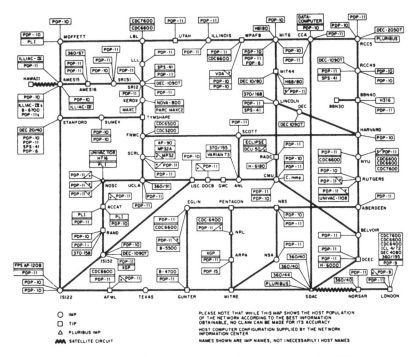

FIGURE 5. ARPANET logical map, 1977. Source: J. A. Payne *ARPANET Host to Host Access and Disengagement Measurements* (Washington, DC: US Department of Commerce, 1978). Digitized by Google. Original from the University of Michigan.

any of its constitutive nodes cease to operate, therefore maintaining the remaining parts of the network operational. The macabre implications behind this design stem from the context of the Cold War between the United States and the Soviet Union, with the escalating threat of nuclear attacks that could decimate entire cities or regions. Given this strategic importance, it was ARPA, the Advanced Research Projects Agency of the US Department of Defense that sponsored the research and development of this network, hence ARPANET.

Underlying the design and implementation of ARPANET there was a specific geography and territorial logic that prioritized the integrity of research, government, and military infrastructure across the US territory. Thus, the locations that became the first nodes in ARPANET were not random but represented a selection of the most advanced computing centers in the country at the time (see Figure 5 and Map 3): a mix of universities (e.g., UCLA, Stanford Research Institute, MIT), government facilities (e.g., NSA, Pentagon), and defense consulting firms (e.g., RAND,

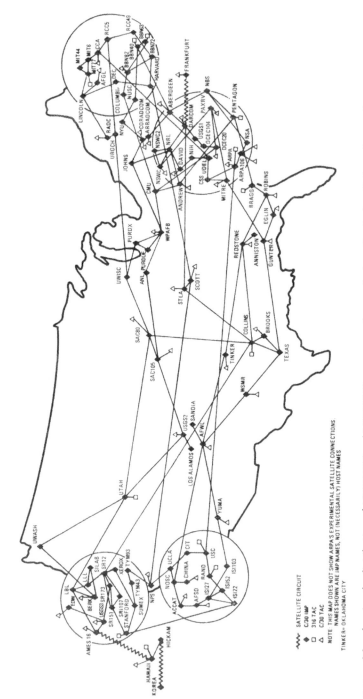

MAP 3. Map showing the distribution of ARPANET and MILNET interconnected networks in 1984. *Author:* Defense Systems Agency.

BBN). Throughout the 1970s and 1980s ARPANET expanded in size and complexity, particularly after its connection to a network created by another part of the US government—the National Science Foundation's Computer Science Network (CSNET), founded in 1981. However, soon afterward NSF established a new network, NSFNET, whose backbone was a series of recently established supercomputing research centers. This led to the decommissioning of ARPANET in 1990, which left NSFNET as the main computer network in the United States, operating for another decade. Throughout the 1990s NSFNET became increasingly open to commercial operators, a process that culminated with this network's decommissioning in 1995, when it gave way to the commercial internet.[27]

Throughout these connections, disconnections, and reconfigurations between computer networks, it is important to highlight their inherent spatiality and its enduring consequences. What we know today as the internet was initially shaped by a geography of research sponsored by the US federal government that responded to territorial defense imperatives in the shadow of catastrophic devastation.[28] Thus, from its inception, the very ideas that gave rise to the internet—the network of networks at the core of the digital economy—were influenced by the necessity of locating digital information throughout a map of key sites and their role in connecting a territory. Upon this geography a specific form of network arrangement (Baran's distributed computer network) was built to ensure the continuous and adaptable flow of information between locations. This infrastructural and geographic reality at the core of the internet should be highlighted precisely because the very construction of the digital economy (as frictionless, instantaneous, "in the cloud"), has thrived on an imaginary of information floating free from geography, untethered from specific locations and the limitations of physical space—even as it is this very tethering that makes the internet (or any communications network) possible in the first place.

The 1982 short story *Burning Chrome* by Canadian science fiction author William Gibson first featured the word *cyberspace*, a neologism that would soon take on a life of its own. However, it was not until its reappearance in *Neuromancer*, the 1984 novel by the same author, that this term gained popularity and found widespread application in reference to networked computer environments. In *Neuromancer* Gibson vividly described cyberspace as a "consensual hallucination experienced daily by billions of legitimate operators, in every nation, by children being taught mathematical concepts. . . . A graphic representation of data abstracted from banks of every computer in the human system. Unthinkable complexity. Lines of light ranged in the nonspace of the mind, clusters and constellations of data. Like city lights, receding."[29] This passage's evocative imagery left a durable cultural imprint that cemented the widespread perception of computer networks as information worlds somehow existing "separate" from, or somehow "outside" physical and social reality. Key to the conceptual construction of cyberspace, Gibson stresses its immaterial, abstract, and aspatial nature ("the nonspace of the mind"). These qualities have since become recurring features in the cultural representations of the internet and other digital environments, which to this day find resonance in how we often conceive of the digital economy, as a "realm apart." However, as powerful as Gibson's imagery was, it gained widespread influence in part because it was released into a receptive cultural context where its vision of cyberspace was swiftly embraced.

In the book *From Counterculture to Cyberculture*, Fred Turner documents how the emergence of the idea of cyberspace was shaped by a particular cultural milieu of the San Francisco Bay Area between the late 1960s and the late 1990s. The local confluence of post-Vietnam-era counterculture, technological utopianism, and later Reagan-era and Silicon Valley–centered forms of fast capitalism served as a catalyst for the notion of an electronic world disconnected from the mores, norms, institutions, and constraints of the physical and social setting of the time—and specifically impervious to the regulations that may have applied in that setting.[30] This sentiment severing cyberspace from physical reality and its governing structures was memorably encapsulated by John Perry Barlow, the ex–Grateful Dead lyricist and cyberlibertarian political activist, in an influential manifesto. Published in 1996 under the title "A Declaration of Independence of Cyberspace," the text provocatively defines cyberspace through a forceful rejection of the institutions and systems of political and economic order organizing the material world. Resulting from this schism, cyberspace then becomes a new "global social space": a world that is "everywhere and nowhere, but it is not where bodies live," and "a civilization of the mind." By rejecting the legitimacy of governments ("You are not welcome among us. You have no sovereignty where we gather"), Barlow provides a justification for a libertarian utopia held together not by the structuring power of the state or other social institutions, but by "transactions, relationships, and thought itself, arrayed like a standing wave in the web of our communications."[31]

The conception of cyberspace as a place "beyond location" became increasingly influential as the internet began to spawn its own culture, complete with its behavioral codes, languages, and forms of social organization. In the minds of many users, digital information flows had transcended the barriers of physical space and made geography, proximity, and location obsolete. For Barlow, this aspatiality was the basis to declare the illegitimacy of governmental intervention, and consequently, any external regulation of the informational flows that made up "cyberspace." For others, the consequences were less grandiose, but perhaps more profound: as the internet grew in popularity and adoption, the "offline" lives of everyday people all over the world began to change to adapt to the new hybrid reality enabled by increasingly immersive, round-the-clock instantaneous digital communications. Yet, regardless of how much early discourses of the internet insisted on its aspatial nature, new technological developments would have to embrace the fundamental role of space to this network's underlying logic and its architecture. For one, the distribution of internet connectivity, and its purported benefits, was highly influenced by factors such as national income, location, class, and degree of urbanization—all of which led to the growing discourse of "the digital divide."[32]

Beyond the spatial disparities in internet access, however, the spatial constitution of the substance of the network itself continued to change. As this network evolved throughout the first decades of the twenty-first century, the content it delivered went from file directories and static web pages to highly interactive and customized experiences to users, which included multimedia streaming services, online video games, social media, and online shopping. Yet, even as these experiences bolstered ever more sophisticated and immersive online environments, seemingly detached from the physical world, they also came to depend more closely on location. Therefore, a spatial paradox undergirded some of the key developments of Web 2.0, like social media, participatory platforms, and customized content: even as digital environments seemed increasingly self-contained due to their interactive, immersive, and social capabilities, these environments became more reliant on identifying, tracking, and monetizing the location of users in myriad ways.

Key to this spatial shift in the development of digital networks, and the emergence of digital capitalism more broadly, was the advent of geolocation, or the identification of the geographic location of devices connected to the internet. There are multiple ways of geolocating information flows on the internet, but two broad categories have gained particular prominence due to their widespread usage and their role in enabling a wide range of commercial activities that fuel the growth of the digital economy. The first method uses IP, or Internet Protocol addresses, which are the unique numerical codes that identify each individual machine through its connection to a specific internet service provider's computer network. For many years IP addresses were not widely understood as geographic identifiers, and their use for location purpose was rare. However, in the past decade and a half, the acknowledgment of the spatial structure underlying the internet enabled increasingly precise location based on the identification of IP numbers. This in turn became an important pivot for the internet in terms of customization of content surveillance, targeted advertising, and many other services that explicitly relied on the segmentation of internet users based on the geography of their IP addresses.[33]

The second method of geolocation that gained importance in the development of the internet is the use of devices with GPS, or Global Positioning System, capabilities.[34] This method relies on a network of satellites orbiting the globe, which can be triangulated to establish the coordinates of receiving devices.[35] While GPS was developed independently from the internet, it gained extensive use and importance with the advent of smartphones in the late 2000s, since these devices began to incorporate GPS receivers, unlike most previous cell phones, or even laptops and desktop computers. As a result, the use of GPS devices for geolocation also came to shape the development of a broader array of products and services that we now experience through the mobile internet such as location-aware social media, navigation services, dating applications, digital platforms, and many others. Paul Ceruzzi succinctly synthesizes the widespread ramifications of the advent of GPS and its integration with an ever-expanding range of devices: "It is not just having GPS on a smartphone or in an automobile that is so revolutionary—it is the integration of GPS with other computer-and-internet-based applications, including cellular networks, digital maps, satellite weather information, and so on."[36]

In the next section of this chapter, I delve into the development and implications of IP geolocation, and specifically its role in structuring the process of value allocation in digital markets. A discussion of GPS geolocation and its implications for digital capitalism is woven throughout the next two chapters. Chapter 4 discusses the satellite infrastructure that enables GPS, highlighting its relationship to the other key use of satellites for geospatial data purposes: Earth observation. In chapter 5 the focus is on the crucial role of GPS in enabling the operations of transportation networking companies, and therefore the emergence of platform mobility. As will become clear, both IP and GPS geolocation methods have been fundamental to the locating of digital information in physical space, and in doing so, they have enabled new forms of valuation and marketization of this information, which are at the core of digital capitalism. The following paragraphs examine how the use of IP geolocation has become crucial for locating, valuing, and marketizing digital information, therefore constituting a core component of the (spatial) architecture of digital capitalism.

ALLOCATION

In the early 2000s, most interactions on the internet took place behind a veil of anonymity. Internet users today, by contrast, routinely disclose sensitive and identifiable information across different platforms and for a variety of purposes, sharing personal information on social media, inputting financial data to make purchases, and even making their location known to third parties while using navigation or ride-hailing applications. Location has become an integral part of how user profiles (and their analytics) are assembled, which makes it essential to the business activities of all manner of companies who use the internet, from online giants like Facebook and Google to platforms like Airbnb and Uber, and even brick-and-mortar retailers like Walmart and Target. Yet this state of affairs is a relatively recent development, sharply contrasting with the way nascent digital markets operated in the early 2000s. In fact, establishing users' identity while safeguarding personal information was one of the key challenges that digital markets had to satisfactorily address to gain a foothold—even if large scale data breaches are still worryingly all too common. This meant that complete anonymity between all parties, which had become an expectation in much of the early internet, became a barrier to establishing trust in transactions, a fundamental element to the construction and operations of markets.[37] Trust, identity, and location are deeply intertwined: some of the key questions at play in any market transaction involved determining whether parties are who they say they are, whether they are located where they say they are located, and whether each party can take the others' word as true in any of these matters. Furthermore, location in transactions is also fundamental to the construction and regulation of markets, particularly from the perspective of the state. Constitutive elements of global capitalism, such as trade treaties, tariff schedules, and taxation of goods and services, follow a territorial logic where specific criteria apply, depending on the location of each element in a complex transaction network: from the buyer and the seller to the site of production, assembly, or transportation trajectory.

The identity-location-transaction link, so central to the workings of capitalism, poses a particularly vexing problem for the notion of "cyberspace" since it throws into question one of its foundational assumptions: that the internet is a realm apart where location in physical space is irrelevant. Under this conception, if cyberspace was a domain governed by its own set of rules, and where states and other territorial powers "of flesh and steel" (in John Perry Barlow's words) had no legitimacy or jurisdiction, then certainly location should play no role in how users transact with each other behind a veil of anonymity. Yet the privatization of the internet in 1995 and its turn toward more intensive commercialization required a resolution of this conundrum that would enable a consistent measure of trust that would, among other things, underpin the construction of digital markets.[38] The resolution of the identity-location link would bring IP geolocation to the forefront, revealing the deep spatial imbrications of the internet, and the irrevocable tethering of digital information flows to location, place, space, and territory—in a word, to geography. The paragraphs below discuss a key court case in the spatialization of the internet. This international dispute, centered around the actions of the early internet giant Yahoo!, brought IP geolocation to the fore, paving the way for the explicitly spatialized construction and governance of digital markets.

UEJF and LICRA v. Yahoo! Inc and Societé Yahoo! France was a court case that began in the year 2000 and mutated into a series of legal battles that continued for over half a decade in courts across France and the United States. While purportedly the case addressed an issue of freedom of expression on the internet, one of its consequences was to bring to light the deeply spatial structure of the network's operations as well as the extent to which governments can claim jurisdiction and control the flows of digital information.[39] The case began when two French organizations, La Ligue Contre Le Racisme et l'Antisémitisme (International League Against Racism and Anti-Semitism, LICRA) and Union des Étudiants Juifs de France (Union of French Jewish Students, UEJF), filed a lawsuit against Yahoo! and Yahoo! France in the Superior Court of Paris. The basis for the lawsuit was that internet users located in France were able to view and purchase Nazi memorabilia on the Yahoo! auction service. While access to, and purchase of such content was not illegal in the United States, the location of Yahoo!'s headquarters, it is however strictly prohibited in France under Article R645-1 of the French Criminal Code.[40] On May 22, 2000, the Superior Court of Paris issued an interim judgment in favor of the plaintiffs, ordering Yahoo! and Yahoo! France to take the following actions, among other measures: (1) Yahoo! Inc. should "destroy any and all computer information held directly or indirectly on its server and correlatively . . . cease all lodging and servicing thereof on the territory of the Republic from the site 'Yahoo.com'" relating to Nazi memorabilia as well as antisemitic content. (2) Yahoo! Inc. and Yahoo France should prevent holocaust denial content being accessible from within "the territory of the French Republic."[41]

Yahoo!'s response brought to the surface the complex geographies of the internet by combining the logic of territorial jurisdiction with the prevailing aspatial imaginary of cyberspace. First, Yahoo! claimed that the location of its servers in US territory meant that the information contained within them was protected by the First Amendment of the United States Constitution, ensuring freedom of speech. Secondly, Yahoo! retorted that they had no way of knowing who might access the auction site, and therefore no way of either making individual postings available to certain specific users or preventing those very users from accessing items that are unlawful in their jurisdiction. Furthermore, they asserted that even if such means of determining users' location and tailoring content to them did exist, deploying them would not only "entail unduly high costs" for Yahoo!, but also "might even place the company in jeopardy," and would even "compromise the existence of the internet as a space of liberty and scarcely receptive to attempts to control and restrict access."[42]

A key vulnerability of Yahoo!'s argument was identified by LICRA's defense council, who countered that the search engine was providing results that were tailored according to the primary language of a users' location. The corollary of this argument presented Yahoo! with a very difficult challenge: If Yahoo! had no way of knowing where users were located (and therefore no way of preventing transactions that might be illegal in said location's territorial jurisdiction), then how could the company provide online content tailored to specific languages and locations? This challenge to Yahoo!'s argument about the (a)spatiality of the internet was supported by the demonstration provided by Cyril Houri, a French software engineer who testified as an expert witness on behalf of the plaintiffs. Houri showed that using geolocation technologies, a novel technique at the time, one could block up to 90 percent of users in France from accessing Yahoo! sites.

This blocking capability was enabled by the very geographical aspects of the internet's information architecture that Yahoo! had downplayed. Specifically, it involved the use of Domain Name System (DNS) data-bases and targeting the content that specific IP addresses can access ac-cordingly. This demonstration opened a discussion that recovered the spatial configuration that has structured the internet's architecture from the beginning. Domain names and internet service providers are all or-ganized spatially since the domain name system "is a naming database in which internet domain names are located and translated into internet Protocol (IP) addresses. The domain name system maps the name people use to locate a website to the IP address that a computer uses to locate that website."[43] This mapping in turn allows geolocation technologies to use DNS databases to make reasonable inferences about where users are located.

While much of the mythology of cyberspace has been premised on severing the link between digital information and physical location, the demonstration of the technical feasibility of "geolocating" users brought to the surface the profoundly geographic nature of the internet and showed the possibility of tailoring content (and hence building digital markets) around it. Bringing to the surface the underlying geography of the internet was not only central to Houri's demonstration in court, but it was also instrumental in the development of a new set of technologies that would enable the explicitly spatialized internet upon which today's digital capitalism is built. On April 4, 2000, only a few weeks before the Superior Court of Paris ruled in favor of the plaintiffs in the Yahoo! case, Houri's company, Infosplit Acquisition LLC, along with Microsoft Technology License LLC, filed a key patent at the US Patent and Trade-mark Office. This patent, number US6665715B1, was named "Method and systems for locating geographical locations of online users,"[44] and it not only introduced a breakthrough innovation in refining the means to geolocate information (and users) on the internet, but it did so by ex-plicitly leveraging the underlying geographic architecture of the internet. Figures 6, 7, and 8 reproduce three of the illustrations contained in the patent filing, which show how this geolocation method makes use of the embedded geography of the internet to establish the geographic loca-tions of users.

At the time, Yahoo! was the leading search engine, and had aggres-sively branched out into many other areas by acquiring companies and by setting up operations in new national markets. One of the consequences

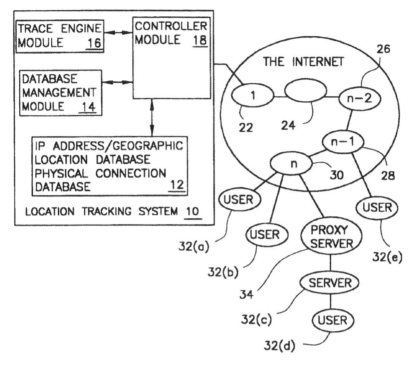

FIGURE 6. Diagram included in patent US6665715B1, "Method and systems for locating geographical locations of online users," United States Patent and Trademark Office filing on April 3, 2000. *Source:* Google Patents. Inventor: Cyril Houri.

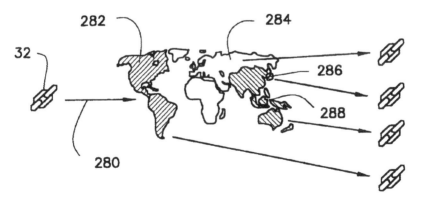

FIGURE 7. Diagram included in patent US6665715B1, "Method and systems for locating geographical locations of online users," United States Patent and Trademark Office filing on April 3, 2000. *Source:* Google Patents. Inventor: Cyril Houri.

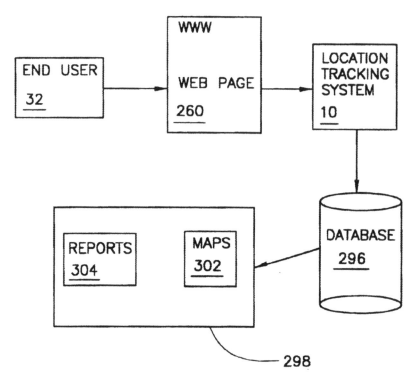

FIGURE 8. Diagram included in patent US6665715B1, "Method and systems for locating geographical locations of online users," United States Patent and Trademark Office filing on April 3, 2000. *Source:* Google Patents. Inventor: Cyril Houri.

of this expansion was that people all over the world were able to access their services—that is, the company's millions of users were distributed throughout many territorial jurisdictions. In a sense, this was not an anomaly for a growing internet company—after all, the prevailing ethos dictated that cyberspace knew no boundaries. However, what the ruling from the Superior Court of Paris demonstrated was that Yahoo!, other tech firms, and the internet as a whole, were irremediably tied to physical space, and consequently beholden to the power of states through territorial jurisdictions. In the words of Jack Goldsmith and Tim Wu, two legal scholars who have studied this case extensively, one of the questions raised by *UEJF and LICRA v Yahoo! and Yahoo France* is, effectively, "who controls the internet?"[45] More specifically, as the focus on geolocation allows us to see, determining the answer to this fundamental question involves also determining the result of who can locate information on the internet, who can connect producers, consumers, and mediators of this information. These processes, in turn, are crucial for defining to what extent "cyberspace" is effectively under the umbrella of states and other territorially bound entities.

While the case between LICRA and Yahoo! took many turns and continued for several years, the issue that took on more salience in those latter stages was no longer that of the location of the users (and therefore the ability to know whether transactions were legal or illegal in certain places), but rather the jurisdiction in which the case should be tried. This is the second aspect of the spatial logic mustered by Yahoo!, which is premised on the principle of territorial jurisdiction. Contrary to claims that states have no right to rule over cyberspace, Yahoo! sought the protections offered by its home jurisdiction (the United States) while claiming that French courts did not have the ability to regulate the company's activities, regardless of where its end users were located. This argument spurred a years-long trans-continental legal saga that spanned the French Superior Court of Paris, the United States District Court for the Northern District of California, in San Jose, and the United States Court of Appeals in the Ninth Circuit. In 2006, the latter court ruled that the dispute could not be adjudicated in the United States by concluding that it had no jurisdiction and could therefore not enforce the orders issued by the French court.

Today Yahoo! is no longer a leading search engine, and much of the internet is vastly different from what it was during these trials in the first years of the century. Yet the questions, methods, techniques, and arguments raised in this multiyear series of transatlantic legal disputes were crucial to determine issues as varied as the grounds for regulation of information in frameworks such as the European Union's General Data Protection Regulation (GDPR) and the Digital Single Market (DSM), the degree to which authoritarian governments can shut off, or control access to the internet within their territories, how much intelligence agencies can track specific users' activities online, and where they can do it, and even the content of the media libraries we can access via streaming services, such as Netflix, who modify their offerings depending on the geographic locations of users due to the specific geography of distribution agreements with studios and content producers.[46]

In the years since the Yahoo! case, the recognition of geolocation and its possibilities has gone from the relatively narrow applications demonstrated by Cyril Houri in the courtroom to a standard feature that enables the valuation and marketization of myriad digital products and services. Whether it is carried out via IP addresses or GPS receivers, geolocation is crucial in establishing a link between digital information flows and geography. In the first case this is enabled by the very infrastructure of the internet, while in the second case it is generated via the combination of satellite constellations and GPS-enabled mobile phones. The result in both cases is the inversion of the cyberscape imaginary of information flows abstracted from physical space. Contrary to that enduring notion, it is through geolocation that the value of digital information is realized in differential ways according to its multiple and complex relationships with space and place: from trace data that sheds light on patterns of user movements to targeted advertisements drawing on user behavior and decades of geodemographic marketing to the creation of geographically differentiated and even individually customized media markets where users across countries, households, and even personal profiles are presented with a different set of offerings. In the next chapter, I discuss how one of the key technologies for the advent of geolocation via GPS, artificial satellites, have also provided other forms of geographic information that in turn present different configurations of location, valuation, and marketization, and with them new kinds of integration into digital capitalism: Earth observation and remote sensing data. Chapter 5 then turns to examine how the incorporation of different kinds of geolocation and other technological developments have been instrumental for the profound transformation of the automobile industry by enabling the emergence of two new forms of mobility, each with its own forms of valuation, marketization, and spatial ramifications: app-enabled ride hailing and autonomous vehicles.

CHAPTER 4

Eyes in the Sky
and the Digital Planet

AN EMERGING SATELLITE ECOSYSTEM

This chapter analyzes the emergence of a new satellite ecosystem over the course of the past decade, focusing on the role of Earth observation and other remote sensing data as forms of geographic information increasingly integrated into the broader workings of digital capitalism, and representing diverse configurations of the *location, valuation, marketization* conceptual triad at the core of this book. The recent and rapid development of small, micro-, and nanosatellites has reignited interest in outer space and led to a booming economy of private firms as well as governments around the world to acquire or expand their satellite capabilities. This new satellite ecosystem presents a marked difference from that which developed within the framework of the Cold War in the second half of the twentieth century. I argue that a key to understanding the rapid expansion of the satellite industry in recent years is to consider its deepening integration with the logics, products, services, and (institutional, corporate, and informational) networks of digital capitalism. While the previous chapter focused on the development of geolocation as a technical capability that established a link between information flows and physical space, in this chapter I expand on the role of Earth observation (EO) imagery and other forms of satellite-based remote sensing as types of geographic information that have become highly valued for their ability to provide synoptic views of sections of the planet across multiple scales with rapidly improving spatial and temporal resolution. Although satellites predate most of today's digital technologies, they can be seen as products of the same general political and technological milieu that gave rise to some of the core infrastructures of digital capitalism—such as the internet, particularly through its origins in ARPANET. As such, satellites are also fundamental to the development and operations of key features of digital capitalism, like GPS, and therefore central to processes of geolocation, valuation, and marketization. At the same time, satellites have received renewed interest due to their multiple synergies with core areas of digital capitalism: from the provision of internet services (via the Starlink constellation) to the application of artificial intelligence to analyze ever-closer-to-real-time observations of the surface of the Earth. The latter has boosted a rapidly growing market on satellite data services and analytics, whose size was estimated by Straits Research at $6.81 billion in 2022 and was expected to grow to $61.22 billion by 2031 through applications across industries such as defense and security, engineering and infrastructure, maritime, energy and power, and transportation and logistics.[1]

A consequence of technological developments in satellites as well as renewed interest in outer space, the collection of Earth imagery has dramatically expanded in recent years (see Map 4). Unlike decades ago, when government programs like the joint NASA/USGS Landsat operated a relatively small number of satellites that formed the core of the global satellite infrastructure, today most EO data is captured by privately owned and operated constellations of up to hundreds of satellites working together. For instance, the San Francisco–based satellite startup Planet (formerly known as Planet Labs), founded in 2010, operates over 200 active satellites. Among these is the PlanetScope constellation, which at the time of writing had over 130 "nanosatellites called Doves weighing only 5.8 kg each." This constellation "provides 3-meter multispectral image resolution for a variety of mapping applications including several humanitarian and environmental applications, from monitoring deforestation and urbanization to improving natural disaster relief, and agricultural yields around the world."[2]

The PlanetScope constellation is a useful entry point to investigate the complex outer space economy that has developed over the past decade and includes not only firms focused on Earth imagery like Planet, but also rocketry, telecommunications, and scientific research, among many other areas. In fact, 48 of the satellites in this constellation were put in orbit by a single, record-breaking launch by another standout firm in the new satellite ecosystem: SpaceX's Falcon 9 rocket in 2021. This rocket carried 143 satellites in total, the highest number placed in orbit in a single launch. In addition to Planet's nanosatellites, the rocket launched out of Cape Canaveral Space Force Station, Florida, also carried satellites by other firms from around the world such as iQPS (Japan), Capella (United States), and Iceye (Finland), as well as 10 of SpaceX's own satellites, which are part of the firm's Starlink mega-constellation of over 2,500 satellites, which aims to deliver global-scale broadband internet coverage.[3]

Previously the record of the most satellites placed in orbit was held by the thirty-ninth mission of the Indian Polar Satellite Launch Vehicle (PSLV-C37), launched by the Indian Space Research Organization (ISRO) in 2017 from the Satish Dhawan Space Centre on Sriharikota Island in Andhra Pradesh, off the Bay of Bengal. This rocket carried 104 satellites on board, the primary of which was India's Cartosat-2 series. According to ISRO, this satellite is equipped with panchromatic and multispectral cameras, sending remote sensing imagery that "will be useful for cartographic applications, urban and rural applications, coastal land use and regulation, utility management like road network monitoring, water distribution, creation of land use maps, change detection to bring out geographical and man-made features and various other Land Information System (LIS) and Geographical Information System (GIS) applications."[4] Aside from Cartosat-2, the remaining 103 were passenger satellites from private firms and governments: 96 from the United States, among which 88 were Dove satellites to be added to Planet's constellation, as well as 8 LEMUR Satellites from another San Francisco–based satellite startup, Spire Global Inc, and which "are meant for providing vessel tracking using Automatic Identification System, besides carrying out weather measurement using GPS Radio Occultation."[5] The remaining satellites were from countries such as Israel, Kazakhstan, the Netherlands, Switzerland, United Arab Emirates, and the United States, as well as India itself.

As these examples suggest, satellite launches and operations today are characterized by the proliferation of private firms, the wide range of instruments, technologies, applications, and the expanding number of countries with some degree of investment in developing, acquiring, or providing satellite capabilities. Altogether, this signifies the emergence of a satellite ecosystem that is markedly different from its superpower-dominated predecessor, which was a core element of the Space Race in the second half of the twentieth century and central to the structuring feature of that time period's geopolitical order: the Cold War.

Throughout this chapter, I examine three key factors necessary to understand the configuration and stakes of this emerging ecosystem, as well as its integration with the broader digital economy.[6] The first of these is the appearance of technological innovations such as progressively miniaturized satellites, cheaper launch vehicles, and AI-powered imagery analytics. The second development comes from the underlying geopolitical rearrangements that have changed the balance of power and the capabilities for satellite technologies and space exploration since the end of the Cold War. This has opened the door for incipient space powers to consolidate and expand their existing presence in orbit (e.g., India, China), while also broadening the field of newcomers seeking to materialize their own satellite capabilities. The latter includes countries like Kenya, Nigeria, Malaysia, Philippines, Bhutan, Egypt, Bangladesh, Costa Rica, Paraguay, Bahrain, Armenia, and Moldova, among many others (see Figure 9). A third development that has fueled the growth of this new satellite ecosystem and reoriented its objectives and operations is the leading involvement of private firms seeking new commercial endeavors in outer space. This increased commercial orientation reflects the realignment between the state and private sectors that has characterized the global process of neoliberalization throughout the past four decades. Building on these three developments, in the latter part of the chapter I examine the various property regimes that organize the access, use, ownership, and commercialization of various satellite imagery platforms. This analysis in turn informs a discussion of how the emerging satellite ecosystem is intertwined with the creation of new markets for remote sensing imagery and other data products ever more closely integrated into the digital economy.

Shifts in the development and commercialization of satellite technologies have unfolded in parallel to a process of intensifying digitization across many other sectors and industries—all fueled by the continuous development of information and communication technologies. Importantly, unlike in prior eras of space exploration and Earth observation, widespread digitization makes it possible for new streams of satellite data and imagery to be readily integrated with other sources of big data, digital information networks, as well as artificial intelligence algorithms and other powerful computational tools. This in turn enables the creation of innovations like queryable databases and analytics applications that allow for the near real-time monitoring of Earth's surface as well as the identification, tracking, and monetization of data on specific places, events, patterns, people, and objects almost anywhere on our planet.

FIGURE 9. CubeSats from the Joint Global Multi-Nation Birds (BIRDS) Project. *Source:* JAXA (Japan Aerospace Exploration Agency).

Among other innovations, the resulting satellite ecosystem has enabled the creation of a "Big Earth Data" economy, where satellite data are collected, used, and recombined with an expanding catalog of commercial and noncommercial technologies and applications—from business analytics, mapping, infrastructure and navigation platforms to environmentally oriented uses like plant and insect biotelemetry, cameras monitoring rainforests, webs of sensors deployed in the world's oceans, and new artificial intelligence tools to analyze pollution data.[7]

In addition to feeding new markets for digital products and services, these new forms of data collection and analysis of the Earth's surface also have political and institutional implications. As Karen Bakker and Max Ritts have argued, a key outcome of these "Smart Earth" technologies is that they "create conditions for significant shifts in environmental governance."[8] Furthermore, Cindy Lin has observed how the use of satellite-collected remote sensing imagery for decision-making purposes can have significant institutional impacts. In the case of state-led forest mapping in Indonesia, she has noted that this practice "shapes political norms of transparency and legitimacy in social institutions."[9] It should be noted that, technological and commercial innovations notwithstanding, the emerging satellite ecosystem continues to feature the government in a privileged—and in many cases commanding—position. Nevertheless, the widening scope of actors, applications, and orientations intermingled in this new satellite ecosystem does mark a qualitative shift from the arrangement established during the second half of the twentieth century, where satellite data and technologies had been primarily developed

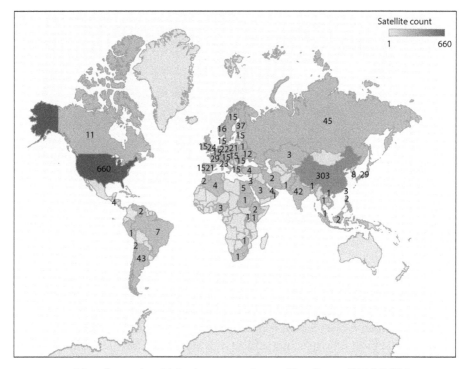

MAP 4. Map of countries with land remote sensing satellites. *Source:* EROS CalVal
Center of Excellence (ECCOE), USGS.

and applied under the purview of government agencies, both civilian and
military, and limited to a small number of countries.

The changes bringing about this new satellite ecosystem are not limited to new technologies or recently collected data but extend retro-
spectively into the past. Renewed interest and investment in satellites,
escalating environmental urgency, demand for data by multiple sectors
of society, and the widespread digitization of information all combine to
simultaneously move the new satellite ecosystem forward while bringing
back earlier developments. A notable instance of this is the recovery of
decades-old data archives for their integration into digital environments,
which endows these data with renewed scientific and potentially commercial value. For example, 2013 marked the thirty-fifth anniversary of
the pathbreaking, if short-lived, Seasat mission. Launched on June 27,
1978, from Vandenberg Air Force Base, in Santa Barbara County, California,[10] this NASA satellite only lasted 105 days in orbit when a "massive short circuit in the spacecraft's electrical system ended the mission

on October 10, 1978."[11] Despite its short lifespan, Seasat made history due to its technological innovations and the first radar images of Earth they produced. According to NASA's Jet Propulsion Lab:

> Seasat's experimental instruments included a synthetic aperture radar (SAR), which provided the first-ever highly detailed radar images of ocean and land surfaces from space; a radar scatterometer, which measured near-surface wind speed and direction; a radar altimeter, which measured ocean surface height, wind speed and wave heights; and a scanning multichannel microwave radiometer that measured atmospheric and ocean data, including wind speeds, sea ice cover, atmospheric water vapor and precipitation, and sea surface temperatures in both clear and cloudy conditions.[12]

In 2013, more than three decades after the launch (and early expiration) of Seasat, engineers at the Alaska Satellite Facility at the University of Alaska Fairbanks Geophysical Institute completed the task of digitizing the data collected by the satellite. The result is a first-in-kind collection of "several thousand unique gray-scale images of the Earth's surface, each displaying 100-square-kilometer sections in high resolution."[13] Importantly, since Seasat was the first SAR satellite, this allows scientists to create a decades-old baseline of data unobstructed by clouds or darkness capturing Earth features like glaciers, volcanoes, and sea ice. The "revival" of Seasat data through digitization encapsulates both the demand for greater volume and variety of Earth observation data, as well as the need for a wide array of technologies to provide it—all of which point to the new pathways opened by the inclusion of the satellite ecosystem into the digital economy. Given these conditions, it should be no surprise that only five years after the digitization of Seasat data the first commercial SAR satellites were launched in orbit, aiming to fill a data gap decades in the making.

The efforts to build a data baseline through the digitization of Seasat's archive point to the rising demand for SAR satellite imagery, given its ability to detect up to millimeter-level changes on the landscape under conditions impossible for other Earth observation satellites, such as cloud cover and darkness. Addressing this demand, Finnish satellite startup Iceye launched the first commercial SAR satellite in 2018, the success of which opened the floodgates to a surge in commercial satellites of this kind. According to an investigation conducted by science journalist Julia Rosen in 2021, "SAR satellites launched by private companies will soon eclipse the number of civil SAR satellites funded by national governments. Overall, nearly 50 are operational, more than double the number in 2018."[14] As of the second quarter of 2022, per the firm's promotional materials, Iceye boasts the largest SAR satellite constellation, having launched over twenty successful satellites since 2018, with over ten additional launches planned in 2022, and additional ones for 2023 and beyond.[15] This increase in commercial SAR launches is contributing to the growth of the satellite ecosystem in general, while simultaneously establishing a dynamic market for these specialized Earth observation products. As reported by Rosen, "Analysts value the SAR market at roughly $4 billion and expect that figure to nearly double over the next 5 years."[16]

The surge in launches, demand, and commercial activity around SAR satellites parallels a broader technological and epistemological shift in Earth observation: from pattern to process. Since their beginnings in 1960 with the first photographic reconnaissance satellite CORONA program,[17] and especially since the Landsat program started in 1972, which became a frame of reference for Earth observation,[18] remote sensing satellites have used different kinds of sensors (from analog to digital cameras to radars and scatterometers) to collect data on the Earth's surface. This has produced an ever-expanding archive of "snapshots" of a world in motion and enabled a better understanding of many of Earth's patterns by scientists, policymakers, militaries, intelligence agencies, NGOs, and, increasingly, private firms. The advances in spatial and temporal resolution in remote sensing data collection brought by each new technological development have caused satellites to move progressively closer to capturing "in real time" the processes continuously unfolding throughout our planet.[19] However, such processes cannot be easily apprehended in their totality via remote sensing, especially given the multiple spatial and temporal scales, and even discontinuous spatialities along which they take place. Floods, vegetation changes, population movements, infrastructure construction or collapse, weather events, and myriad others may unfold over anywhere from minutes to hours, days, weeks, and even years. And, while remote sensing satellites may give us a window into a part of these processes, such a vantage point may also obscure myriad elements that cannot be easily discerned from outer space—sophisticated as the sensors may be. Hence, remote sensing should be seen as complementary to other perspectives (from ethnography to citizen science), rather than substituting for them. Furthermore, we should resist the temptation to consider remote sensing as a "view from nowhere," and instead acknowledge that satellites, as well as the firms and governments that fund them and the people who operate them, are all situated in specific places and social contexts, influenced by material conditions as well as ideological ones. As scholars in the emerging subfield of *critical remote sensing* have argued, to better understand the view of the world provided by satellites, and even transform it, we must "humble and de-universalize [remote sensing] technology, exposing its inescapably earthly origins and its entanglement with particular social, political, and technical conditions and power relations."[20]

Despite the inherent limitations of a top-down, spaceborne perspective, satellites continue to produce ever finer, more numerous, and more frequent "snapshots" of our planet's surface, which can then be used to make incrementally better inferences of processes derived from static patterns. Thus, while a real-time dynamic understanding of Earth's processes in their full complexity may continue to elude satellite remote sensing, this aspiration has motivated and accelerated the ongoing miniaturization of satellites, improvements in their temporal and spatial resolution, their interlinkage into growing constellations, and their integration into myriad other networked digital technologies and applications. It has also motivated renewed commercial ambitions in the expanding outer space private industry. It is no accident that Planet, a satellite startup led by a former NASA engineer, and a key player in this realm, has adopted as its mission statement the goal "to image the entire Earth every day and make global change visible, accessible, and actionable."[21] However, for as much as new satellite technologies are opening the way for the commercialization of outer space, they do so under conditions largely determined by states and influenced by past and present political agendas and geopolitical orders at scales from the national to the global. In the next section I discuss some of these key forces and how they have shaped the conditions that gave rise to the current satellite ecosystem.

THE CHANGING MAP OF OUTER SPACE

Upon the 1957 launch of Sputnik 1 by the USSR, satellites became a symbol of technical prowess as well as geopolitical and economic superiority. Thus, the opening salvo of the Space Race was also a catalyst of the Cold War beyond Earth's atmosphere. Launching and successfully placing a satellite in orbit signaled an unprecedented level of scientific achievement and command over complex technical and organizational infrastructures. It also provided those with control over orbiting satellites newfound communicative and surveillance powers with thinly veiled strategic and military implications. Given the role of satellites as proxy devices at the intersection of multiple key realms, it is no coincidence that they were seen by the two competing superpowers as instruments of power and influence over a future defined increasingly in terms of extra-planetary dominance. The scientific, governmental, and military functions of satellites were thoroughly intertwined from the beginning. On the one hand, Sputnik "made it possible to test satellite pressurisation, to study radio wave transmission and the density of the atmosphere, and allowed scientists to learn how to track objects in orbit."[22] On the other hand, this small metal sphere that orbited the Earth for only three months had both immediate and lasting impacts on the global balance of power, accelerating the ongoing scientific and arms race between the United States and the Soviet Union. One of the immediate consequences was the aforementioned CORONA spy satellite program, jointly coordinated between the United States Department of Defense and the Central Intelligence Agency, which was a direct response to newly acquired Soviet satellite capabilities. In addition to this, Sputnik's success also catalyzed the immediate review of the defense systems of the United States, and the passage of the National Defense Education Act—both moves to retain an edge in the military and scientific competition with the Soviet Union.

These links between military, intelligence, and scientific applications were not unwarranted, since Sputnik and later satellites very evidently combined and enabled advances in these realms simultaneously. In fact, Sputnik was only able to reach orbit because of the advances in rocketry made by the Soviet Union, which were directly related to the development of Intercontinental Ballistic Missiles (ICBMs). For instance, the Sputnik rocket was adapted from the R-7 Semyorka ICBM (Figure 10), which was the most powerful rocket in the world at the time and the first ICBM.[23] In turn, efforts to determine Sputnik's orbit via the Doppler effect of the orbiting satellite allowed scientists at Johns Hopkins Applied Physics Laboratory to develop the TRANSIT system, which was used to establish the position of the Polaris submarine nuclear missile system, one leg of the US nuclear triad (the other two being long-range bombers and ICBMs). The TRANSIT system, considered a forerunner to GPS, became operational in 1967 and was in continuous service until 1996.[24]

The roots of our present-day satellite ecosystem cannot, then, be disentangled from the bellicose ground on which they grew. As has been thoroughly documented, the technological and scientific backbone of the United States' space program was built in the aftermath of World War II with the crucial intervention of Nazi scientists captured, relocated, and employed by the US government as part of Operation Paperclip. The

FIGURE 10. Soviet ICBM R-7 Semyorka. Photo by Sergei Arssenev. Wikimedia Commons.

group of more than fifteen hundred scientists, technicians, engineers, and other personnel included most notably Wernher von Braun, and other key figures of the Nazi German rocketry program,[25] many of whom received leadership positions and top honors at NASA: in 1969 von Braun was joined by transplanted Nazi scientists Kurt Debus, Eberhard Rees, and Arthur Rudolph in receiving the Administration's highest award, the NASA Distinguished Service Medal.[26]

The technical advances made by von Braun and his colleagues before and throughout the war included the Vergeltungswaffe 2, or Vengeance Weapon 2, most commonly known as the V-2 rocket, the world's first long-range guided ballistic missile. Manufactured by Mittelwerk GMBH and built using slave labor from concentration camp Dora (also known as Mittelbau or Mittelbau-Dora),[27] this weapon took the lives of an estimated twenty thousand forced laborers, and thousands more civilians and military personnel in its primary target cities of London, Norwich, Paris, Lille, Antwerp, and Liège during Germany's bombing campaigns.[28] With the transfer of many of this weapon's key architects to the United States after 1945, the V-2 also would come to serve as the foundation for rocketry advances in this country, propelling the American space program past the Soviets in the Space Race—most decisively through Project Apollo and the Saturn rocket series, the fifth of which remains to this day the only launch vehicle to have taken humans to the Moon.

Following the path set by Saturn V more than half a century earlier, NASA has scheduled the launch of Artemis I for September 2022. This mission will have the "the most powerful rocket in the world and fly farther than any spacecraft built for humans has ever flown."[29] Artemis I is designed to lay the foundation for humans to return to the Moon for the first time since Gene Cernan and Harrison Schmitt set foot (and wheels) on the lunar surface on December 11, 1972, as part of the Apollo 17 mission. With Artemis I the US government decisively reenters a rekindled Space Race now characterized not (only) by two superpowers, but by an expanded field of competitors, both public and private, domestic as well as international, and new narratives about how humanity should engage with outer space. In this context, NASA frames the Artemis missions as the incarnation of multiple overlapping scientific, social, and economic goals: "We're going back to the Moon for scientific discovery, economic benefits, and inspiration for a new generation of explorers: the Artemis Generation. While maintaining American leadership in exploration, we will build a global alliance and explore deep space for the benefit of all."[30]

The ambitions of Artemis I are expressed not only in national terms aligned with the United States' geopolitical and broader commercial objectives, but coated in a planetary narrative, or as phrased in capital letters on the project website, "OUR SUCCESS WILL CHANGE THE WORLD."[31] These grand ambitions are at the same time linked to specific societal demands that emerge from contemporary discourse and politics in the United States as well as other countries. For instance, core to this mission's identity is the diversification of participation in space exploration across traditionally excluded social groups, promising to "land the first woman and first person of color on the Moon."[32] This social endeavor is in turn buttressed by the intertwined potential of commercial partnerships and international alliances propelling this undertaking of interplanetary proportions and unmistakable geopolitical symbolism: "[NASA] will collaborate with commercial and international partners and establish the first long-term presence on the Moon. Then, we will use what we learn on and around the Moon to take the next giant leap: sending the first astronauts to Mars."[33] The narrative built around Artemis, combined with renewed interest in space exploration, can help explain the place of NASA in an increasingly marketized outer space economy that is today dominated by private firms. Instead of the primary institutional vehicle to concentrate a nation's space-faring efforts, as it was during the Cold War, NASA is today increasingly understood as a catalyst for new markets through the assignment of projects to contractors, creating new demand, and promoting technological innovation. While economic spillovers have always been important to justify the enormous expenses of NASA's missions, this rationale becomes even more salient in a space economy characterized not by public projects, but by commercial enterprises. This is prominently stated in NASA's website dedicated to answering the question "What does NASA do for you?" which it does in the following terms: "NASA's unique mission provides benefits in big and small ways. Dollars spent for space exploration create jobs, jumpstart businesses, and grow the economy. Our innovations improve daily life, advance medical research, support disaster response, and more. We're constantly evolving and finding new ways to add value."[34] The specific justification for Artemis is therefore also framed under these terms, which stress the commercial potential of space exploration and economic possibilities enabled by the construction of a new space economy: "Artemis I is foundational to the space economy, fueling new industries and technologies, supporting job growth, and furthering the demand for a highly skilled work force. Men and women in all fifty states are hard at work building the Deep Space Exploration Systems to support missions to deep space. NASA prime contractors, Aerojet Rocketdyne, Boeing, Jacobs, Lockheed Martin, and Northrop Grumman currently have over 3,200 suppliers contributing to the milestone achievement that heralds the success of America's human spaceflight program."[35]

Framed in these terms, Artemis should be seen not only as a corner-stone of the revitalized US Space Program in the context of increased competition with other space powers, but as a pivot toward a new environment of competition and cooperation that includes many firms in the private sector that have in fact taken the lead in space exploration during the past decade. Hence, the Artemis mission should also be understood in terms of a state response to the private ambitions for a crewed mission to Mars. This has been expressed most forcefully by SpaceX's CEO Elon Musk, who has pronounced the company's planned Mars mission as the way to turn humanity into an interplanetary species, starting with a million-person colony.[36] Both state-sponsored and private efforts to expand the satellite ecosystem, and beyond it, the outer space economy, are closely interrelated endeavors with vast ramifications across the multiple sectors of the United States, from government and universities to the aeronautic, military, and information technology industries. These developments in turn are related to the rising space ambitions being realized in many countries across the world—perhaps none more significant than China.

As a rising space power, China has crafted a particularly dynamic approach to expanding the field of actors in the satellite ecosystem. Part of this strategy has involved developing partnerships with countries in regions where China intensified its pursuit of strategic interests. The list of countries with satellites launched from Chinese territory includes Brazil, Ecuador, Bolivia, and Venezuela in Latin America; Nigeria and Algeria in West and North Africa, respectively; Sri Lanka and Pakistan in South Asia; Laos and Indonesia in Southeast Asia; and Belarus in Eastern Europe.[37] For instance, all of Venezuela's satellites were designed and manufactured in partnership with Chinese scientists and institutions and were launched from facilities within Chinese territory: the *Simón Bolívar* (2008), or Venesat-1, which provided television and broadband services, and the remote sensing satellites VRSS 1 (2012) and VRSS 2 (2017). *Simón Bolívar* was the product of joint work between Venezuelan engineers, many of whom trained at the Beijing University of Aeronautics and Astronautics, and Chinese personnel at the Chinese Academy of Space and Technology and CGWIC, a subsidiary of the state-owned China Aerospace Science and Technology Corporation. While this satellite was lost in March 2020 after almost twelve years in orbit, its development and launch marked an important moment of geopolitical and technological cooperation that is illustrative of the new satellite ecosystem. *Simón Bolívar* was "CGWIC's first satellite in-orbit delivery contract signed with a Latin American customer, and also marks the first space cooperation project between China and Venezuela."[38]

Serving both military and government communications, this project involved technological transfer between both countries, and was part of a broader cooperation framework that also included the South American country's purchase of military technology from China and "various agreements in the oil industry," Venezuela's main export.[39] While the launch of a satellite is no small endeavor, the context in which it is embedded can help explain its broader significance for economic, technological, and geopolitical relations. For Venezuela, this project was part of a strategy to build telecommunications sovereignty, using the satellite to connect schools, hospitals, military, police, and government facilities primarily in rural areas, serving more than eight million people—nearly a third of the country's population.[40] These broader implications help explain why *Simón Bolívar* was priority for then president Hugo Chávez, who personally championed the project, calling the satellite a development "of strategic and historical importance for Venezuela and China."[41]

Continuing the partnership forged by *Simón Bolívar*, China and Venezuela also cooperated in the development and launch of two subsequent satellites specialized in Earth observation: VRSS 1 launched in 2012,[42] and VRSS 2 launched in 2017.[43] Both satellites are part of a remote sensing program involving the Venezuelan government, the Chinese government, and Chinese state-owned enterprises. Like *Simón Bolívar*, the VRSS-series satellites represent a milestone in the expansion of China's range of action in the space sector while deepening its cooperation with Venezuela, particularly since "VRSS 1 is the first remote sensing satellite China has delivered in orbit to an international customer."[44] The satellite projects with Venezuela are but one example of China's increasingly active space sector, which spans countries in many regions of the world as well as partnerships involving a wide array of economic activities and industries. In that sense, such partnerships can be understood as part of the larger strategic and infrastructural objectives of Chinese foreign policy and the economic expansion beyond its own borders, such as the Belt and Road Initiative.

Though inclusive, given the growing number of participants, as evidenced by China's own sprawling network of space-oriented partnerships, the new satellite ecosystem continues to be structured through power asymmetries, both old and new. Even as innovations like miniature satellites, reusable rockets, and the growing participation of private firms have significantly reshaped the configuration of this ecosystem, many of the forces that gave rise to it during the Cold War endure, as do their geographic expressions. For instance, some of the same colonial relations that drew the map of satellite infrastructure in previous decades continue to underpin the emergence of the present ecosystem: from the secretive Luigi Broglio Space Center, located in Kenyan territory but under the continued (and controversial) control of the Italian government[45] to the unironically nicknamed "Europe's Space Center" located in France's South American overseas department of French Guiana—and from where the historic launch of the James Webb Space Telescope took place in 2022.[46] However, such geopolitical forces are today enmeshed in a much more complex web of relationships involving not only states, but firms and other nonstate actors. The prevailing logic in these relationships is much more explicitly centered around the construction of markets in outer space and their integration with digital capitalism. This new environment changes not only the balance of power between states and other actors, like private firms, but creates greater instability where rapid technological changes and continuous integration of satellites with other industries alters the very notion of satellite capabilities, their applications, and wider implications.

MAKING MARKETS IN (OUTER) SPACE

The satellite ecosystem, while enabling essential scientific, infrastructural, and governance functions, among others, is inseparable from visions of outer space exploration and extra-planetary hegemony such as those discussed above. Intertwined with grand geopolitical aims, settler colonialism beyond Earth, the continued fortification of the military-industrial complex, and narratives of scientific and societal progress, these visions form the backdrop of a renewed interest in the outer space economy—one inextricably tied to, but substantively different from the state-centric superpower competition that characterized the original Space Race. Out of the many sectors and industries involved, satellites have experienced notable dynamism, since their development is more attainable, and their applications more immediate and more versatile than some of the expansive space exploration projects discussed above.

With a drastic increase in the number of satellites, decreases in their cost and size, and technological improvements in the collection of Earth imagery, the commercial potential of satellite data has grown significantly in recent years, attracting much attention and investment. As of 2021 at least ten thousand companies, five thousand investors, 150 Research and

Development Hubs and Associations, and 130 governments were part of what the SpaceTech Analytics research firm calls the "Global Spacetech Ecosystem."[47] This ecosystem was not only resilient during the first year of the COVID-19 global pandemic, but many of the firms within it experienced significant growth. According to a SpaceTech Analytics report for the second quarter of 2021, "Despite the crisis and dramatic fall in companies' capitalization in February 2020, capitalization of 376 publicly traded companies grew from \$3,410B on February 3, 2020, to \$4,030B on March 18, 2021."[48]

Led by the proliferation of private firms dedicated to satellite design and manufacture, this boom in satellites has benefited especially from satellite miniaturization. Thus, the emergence of small satellites, or SmallSats, has become the dominant technological paradigm in the span of less than two decades. According to NASA's definition, SmallSats are "spacecraft with a mass less than 180 kilograms and about the size of a large kitchen fridge." This classification is differentiated into minisatellites (100–180 kilograms), microsatellites (10–100 kilograms), nanosatellites (1–10 kilograms), picosatellites (0.01–1 kilograms), and femtosatellites (0.001–0.01 kilograms).[49]

As a technological paradigm, satellite miniaturization has lowered the barriers to entry, drastically altering who can participate in the satellite business, what satellites are used for, and what satellite infrastructures look like. As suggested above, an important effect has been to allow new competitors, like IT companies and startups, to design, manufacture, and launch their own satellites, and even develop new products and services based on them. Further, it is now possible and increasingly common to deploy constellations of satellites that operate in concert and provide greater coverage (of data collection as well as telecommunications) across the surface of the Earth. In aggregate the changes constitute both a qualitative and quantitative shift that not only expands the satellite ecosystem but also deepens its linkages with the rest of digital capitalism. Evidence of this is how satellite networks today constitute a source of new data and informational products. Simultaneously the satellite ecosystem is rapidly becoming an infrastructure that both complements and enables the internet, the Internet of Things, smart grids and cities, and other networks that interconnect to form the core of digital capitalism.

According to the Union of Concerned Scientists (UCS) Satellite Database, as of January 1, 2023, there are 6,718 satellites orbiting the Earth (Figure 11). Even though there are over a hundred countries with

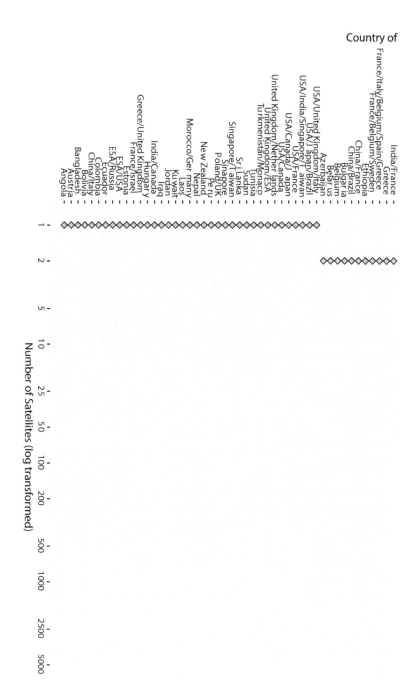

FIGURE 11. Chart showing worldwide distribution of satellites. Elaborated by the author with data from the Union of Concerned Scientists Satellite Database.

Owner or Operator

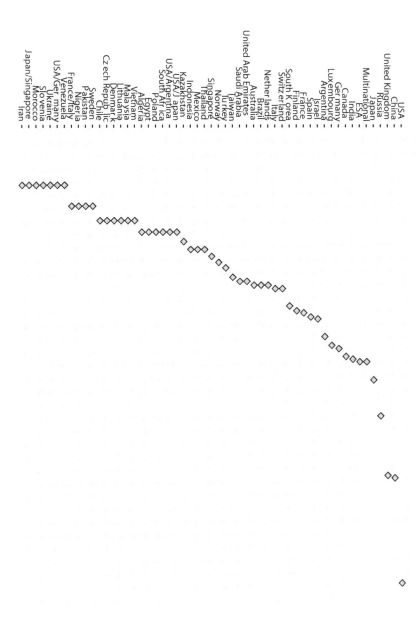

satellites, space-launching capabilities are far from widespread, limited to the countries like the United States, China, Japan, Russia, India, Iran, New Zealand, and members of the European Space Agency. By the same token, most satellites are operated by governments or firms linked to only a handful of countries. The United States has the most satellites with 4,530, followed by China with 592, the United Kingdom with 566, and Russia with 178. All other countries combined have 852 satellites in total.

The satellite ecosystem is continuously changing, with new satellites being placed in orbit, and older ones being decommissioned or becoming nonfunctional. However, in the last decade this ecosystem has followed a trajectory of accelerated growth. For instance, the April 30, 2018, update of the UCS database showed a total of 1,886 satellites around Earth's orbit. Out of these, 859 were assigned to the United States, 250 to China, and 146 to Russia. The rest of the satellites were divided among ninety-five other countries with some form of satellite presence (though few of them had launch capabilities). These numbers show a dramatic enlargement of the satellite ecosystem, with the total number of satellites growing by over 350 percent, and the leading country, the United States, increasing its satellite capabilities more than fivefold in the span of less than half a decade. In second place, China also drastically expanded its orbital presence, more than doubling its number of satellites, while Russia's increased by less than a fifth.

International participation in the satellite ecosystem has gradually increased since the initial status quo defined at the height of the Cold War by the United States and the Soviet Union. From the 1960s through the 1980s a handful of developed countries in Europe and emerging economies in Asia acquired some form of satellite presence, if not necessarily launch capabilities: Canada and the UK in 1962, France in 1965, Australia in 1967, West Germany in 1969, Japan and China in 1970, India and the European Space Agency in 1975, and Israel in 1988, to name a few. Since the end of the Cold War, while many domestic space programs—notably NASA's—have suffered from diminished funding, the opposite has happened with the number of satellites and the share of countries with satellite presence, both of which have skyrocketed. This signals an important shift in the geopolitics and the political economy of outer space, where the growing number of participants, both public and private, mirrors an increased diversification in the range of applications and the integration of satellite technologies with many sectors of the economy. This new field of action has also seen the emergence of new alliances that, though often scientific in the most immediate sense, signal broader patterns of technological, geopolitical, and economic cooperation.

In the realm of Earth observation, which is one of the areas where the satellite economy has experienced the most growth, a crucial component enabling the formation of markets has been the establishment of what I have elsewhere called information policy regimes: legal, institutional, and regulatory frameworks to shape the conditions of collection, use, ownership, and commercialization of information and derived products, often within specific jurisdictions.[50] These regimes often emerge through state action such as regulation, but they are also the product of dynamic interactions and contestations between private firms, civil society groups, international organizations, and foreign countries, as well as differentiated institutional actors within the state itself—at all scales from the national to the local. The balance between public good and private imperatives is one of the key tensions that these regimes must navigate, and in turn shapes core dynamics of any information market, such as who produces the data, who can use it, who can appropriate it, and who can make secondary products from it. In an era of expanding commercial participation from private satellite providers as well as governments from all over the world, and hence new sources of remotely sensed data, analytics, and other derived products, it is worth considering the information policy regimes that have shaped previous satellite programs and products. Below I briefly discuss key transitional moments in the information policy regime that has governed Landsat, the Earth observation satellite program that has set the standard for half a century.

Originally named Earth Resources Technology Satellite (ERTS), Landsat (Figure 12) was funded by the US public and run jointly since 1972 by NASA and USGS, though its operation was transferred for several years to NOAA in the early 1980s.[51] Today the Landsat program "represents the world's longest continuously acquired collection of spaced-based [sic], moderate resolution, land remote sensing data. Four decades of Landsat imagery provide a unique resource for those who work in agriculture, geology, forestry, regional planning, education, mapping, and global change research."[52] As a publicly funded program, one of the core missions of Landsat has been the distribution of remote sensing data not only to government actors, but to the scientific community and the public at large. However, this objective has at times faced vigorous opposition. One such period was the sustained push toward privatization of many government functions that began during the Carter administration and intensified during the Reagan administration. This privatizing

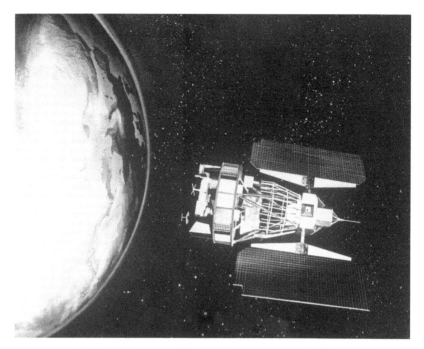

FIGURE 12. Landsat 1 over Lake Michigan, 1978. *Author:* National Aeronautics and Space Administration. John H. Glenn Research Center at Lewis Field.

ethos led President Carter to issue Presidential Directive NSC-54 on November 16, 1979. Through this order, "The National Oceanic and Atmospheric Administration (NOAA) of the Department of Commerce is assigned the management responsibility for civil operational land remote sensing activities in addition to its ongoing atmospheric and oceanic responsibilities."[53] More importantly, the directive stated that the administration's ultimate objective regarding Landsat and other remote sensing infrastructure of the United States was to privatize it: "Our goal is the eventual operation by the private sector of our civil land remote sensing activities."[54]

After the first three satellites of the Landsat program were operational, the privatizing objectives of NSC-54 were realized with the passage of the Land Remote Sensing Commercialization Act of 1984. Soon after, the program was transferred to a private corporation in 1985. Accordingly, the Earth Observation Satellite Company (EOSAT), a joint venture of RCA and Hughes Aircraft Company, received control of Landsat for a period of ten years. During this time EOSAT "operated Landsat 4 and

5, had exclusive right to market Landsat data, and was to build Landsat 6 and 7."[55] The commercial orientation of Landsat's new administration brought change in important ways to the availability of satellite data, as well as the secondary applications to which it could be incorporated. One of the effects was raising the barriers to entry, which left only a few actors with deep enough pockets to purchase the data. As suggested by Emery and Camps, the result was antithetical to the environmental goals that first motivated the creation of Landsat itself: "It should be noted that in an effort to recover their costs EOSAT charges for the Landsat images [were] very high and basically cut out the majority of science users other than oil companies and the insurance industry."[56]

Raising the cost of access thus failed to ensure the continuation of the program, and over the course of the decade in which Landsat was operated by EOSAT, there were continuous funding issues. A point of inflection in the Landsat's trajectory came in 1989, when Congress did not appropriate the necessary funds for the continuation of the program beyond six months. This caused NOAA to direct EOSAT to shut down satellites Landsat 4 and Landsat 5, threatening not only the collection of vital environmental data, but potentially "the jobs of thousands of space workers worldwide."[57] This dire situation brought into stark relief the divergence between the priorities of the private corporation running Landsat and the public interest. As noted at the time by Thomas Pyke, NOAA associate administrator, "the company [EOSAT] stands to earn as much as $10 million from the sale of Landsat archives pictures regardless of the fate of the satellites."[58] This crisis was averted at the last minute when Vice President Dan Quayle, who headed the newly created National Space Council moved to secure emergency funding.[59] However, while this intervention gave Landsat a new lease on life, it did not last long. The same scenario that had led to the need for emergency funding was repeated multiple times in the following years. This situation came to a head in 1992, when "various efforts [were] made to procure funding for follow-on Landsats and continued operations, but by the end of the year EOSAT ceased processing Landsat data."[60]

After years of continued instability that imperiled the United States civil remote sensing infrastructure, in 1992 Congress passed Public Law 102-555, the Land Remote Sensing Policy Act. The objective of this Act was "to enable the United States to maintain its leadership in land remote sensing by providing data continuity for the Landsat program, to establish a new national land remote sensing policy, and for other purposes."[61] One of the effects of this Act was to ensure the continued collection and availability of satellite remote sensing data by the US government, while clearly delineating the creation of a market as a sphere for private actors. As a recent study by the National Geospatial Advisory Committee reported, "While some worried that free Landsat data would interfere with the private sector, the opposite has often been the case. Free availability of Landsat data has allowed the number of users and applications for land remote sensing to increase significantly, thus massively expanding the potential market for commercial remote sensing data."[62]

The Landsat saga can be read as a parable for the many ways in which an information policy regime can shape the contours and operations of an information market as well as the provision of a public good. While the Land Remote Sensing Commercialization Act of 1984 strove to commercialize satellite data collected by Landsat directly, the Land Remote Sensing Policy Act of 1992 restrained that impetus and ensured the role of government as a provider of informational inputs, leaving the commercialization activities to the private sector. With a recent political and technological environment more explicitly aimed at fostering the development of a commercial satellite industry, it remains to be seen how enduring tensions between privatization and the creation of public goods will influence the current satellite ecosystem's deepening integration with other domains of digital capitalism. This question takes on renewed importance when we contrast the increasingly privatized nature of the current satellite ecosystem with the context in which Landsat's information policy regime was developed, when the government had a near monopoly on the collection and distribution of satellite data. From a public interest perspective, ensuring the widely available access of satellite data at little or no cost (today accessible through digital platforms) has been an enormous benefit for the scientific community and society at large. It has also catalyzed the development of many secondary applications that use satellite data, such as the mapping platforms covered in chapter 2 and the development of GPS. The collection and distribution of satellite data places the government as a provider of essential infrastructure as well as informational inputs necessary to build a digital economy. While the state of the Earth observation market is still in its relative infancy (despite the boom in new satellites and the proliferation of data), we can learn from a prior policy decision that had the effect of making GPS location data widely available, which had deep repercussions in the digital economy—some of which are discussed in the next chapter.

This decision took effect on May 1, 2000, at 8:00 p.m. EDT, when the Clinton administration unscrambled the signals emitted by the Global Positioning System (GPS) satellites.[63] This order meant that the US military removed the intentional noise added to satellite-enabled GPS signals used by civilians, which had the effect of increasing locational accuracy tenfold. Such a dramatic improvement spurred a wave of location services such as car navigation systems and later smartphones equipped with GPS. The proliferation of these location-enabled technologies would in turn serve as a springboard for increasingly sophisticated digital applications and services in the early decades of the twenty-first century. The next chapter addresses a category of applications that are only possible due to the positioning accuracy made available through GPS: the mobility revolution brought about by the ride-hailing economy, along with other new forms of transportation such as autonomous vehicles.

CHAPTER 5

People, Platforms, and Robots on the Move

NEW MOBILITY PARADIGMS

This chapter develops a geospatial perspective to examine the ongoing transformation of mobility associated with two recent developments: ride-hailing platforms and autonomous vehicles (AVs). By foregrounding the geospatial foundations of these developments (such as geographic data collection, real-time mapping, and navigation technologies), the chapter shows how ride hailing, AVs, and other recent innovations in mobility generate new relationships between the location of data, the valuation of goods and services, and the construction of new markets around emerging forms of mobility. This in turn both reproduces existing spatialities of digital capitalism, and creates new ones, from the co-location of drivers and passengers in specific spots in the city due to the affordances of ride-hailing apps to the rewriting of traffic rules and governance of transportation infrastructure as a response to AVs. Considering the variety of geospatial tools and the structuring roles they play in new mobility technologies; it is necessary to establish the connections between these elements as well as how they all work collectively to expand the scope and spaces of digital capitalism.

Transportation network companies (TNCs) like Uber, Lyft, or DiDi Chuxing operate ride-hailing platforms that connect drivers with passengers through smartphone applications. These are core to the broader "platform economy" that has dramatically expanded in the past decade to provide services on everything from home rentals to dog walking, freelance labor, grocery, and food delivery—as discussed in chapter 1. The key characteristic of ride-hailing platforms is their position as intermediaries with the technical capabilities to coordinate the matching of passengers seeking rides and drivers offering them.[1] Through coordination and matching, TNCs and their platforms command a position of power since they collect and centralize information from all parties; set the prices, fees, and payments; and internalize all key components of each transaction. The daily operations of these platforms rely on a geospatial infrastructure that can accurately identify, pinpoint, and track in real time the location and geographic distribution of passengers, drivers, routes, and traffic. This seemingly straightforward matching and coordination process is at the heart of a global upheaval in mobility, particularly concentrated in urban areas. As is now evident in thousands of cities around the world, ride hailing has become a very popular form of transportation, and a major industry encompassing a collection of billion-dollar TNCs based in a growing number of countries and with operations in many more—from US-based firms like Uber and Lyft to DiDi Chuxing (China), Grab (Singapore), Gojek (Indonesia), Gett (Israel/UK), and Careem (Dubai). These companies lead a market of substantial size and global scope. According to a recent *Research and Markets* report, the ride-hailing market is currently valued at $62.1 billion, with Asia Pacific as the leading region, followed by Western Europe. The same report notes that even after the contractions precipitated by the COVID-19 pandemic, the ride-hailing market grew a substantial 14.8 percent from 2021 to 2022 and is expected to reach $106.19 billion by 2026.[2] While there are forces that may slow down the growth of the ride-hailing market (such as regulation, macroeconomic trends, and changing passenger habits after the COVID-19 pandemic), these projections provide a sense of the economic magnitude and continued growth expectations placed on this form of mobility. In view of their systemic importance for digital capitalism, and increasingly for mobility worldwide, it is therefore essential to understand the role that TNCs' ride-hailing platforms play in rearticulating spatial (and social) relations.

The widespread adoption of ride-hailing platforms suggests that they provide a service that addresses the mobility needs of billions of people across a variety of geographic contexts. However, ride-hailing has also brought with it enormous challenges for the spatial and economic organization of societies as well as their governance, particularly in urban environments where these services are most concentrated. For passengers, platforms like Uber and Lyft promise a smooth mobility experience coordinated via an app, with shorter wait times and increased transparency, safety, and security. For drivers, they promise the freedom of entrepreneurship along with flexibility and a steady income without "working for someone else."[3] On the other hand, even as millions of people have taken up driving for TNCs as a source of income, and millions more opt for ride-hailing apps as their preferred means of transportation, these services have had pernicious effects on the very places that sustain them. For instance, TNCs have been linked to a host of negative impacts, particularly in cities, such as increases in vehicular traffic and air pollution, the underuse and underfunding of public transportation, the deepened precarity of labor, and illegal corporate and lobbying practices in many jurisdictions around the world.[4]

For good or bad, today ride hailing—as a form of transportation as well as a powerful idea about mobility—has become thoroughly embedded in the socio-spatial structure of cities, and to a lesser extent, extra-urban spaces like rural areas. Due to this embeddedness, the task of addressing the geolocational components that allow ride-hailing applications to track and match passengers and riders in a continuously mapped physical space cannot be separated from examining the logics and strategies deployed by the very companies that run these applications, as well as the broader socioeconomic transformations associated with the rise of the ride-hailing mobility paradigm. This requires explaining how the spatial technologies underpinning ride hailing come together to shape new forms of mobility while documenting how all of this is intertwined with the spatial structure of digital capitalism. In the broader scope of this book, connecting the operations of TNCs with the workings of digital platforms to reshape mobility will provide a window into how geospatial tools and technologies embedded in such platforms enable the creation of new transportation markets, and with them, new spaces for digital capitalism.

Beyond the growing footprint of ride hailing, though not entirely apart from it, another technological innovation has attracted substantial public attention and investment, catalyzing policy debates in recent years, and promising an even more automated mobility future: autonomous vehicles. While AVs have been in development for years and remain mostly in testing phase, the degree of hype around them and the very potential for their successful adoption have already prompted tangible transformations in the global system of automobility.[5] A key aspect of AVs is how, without having been deployed or adopted at scale, they have catalyzed a large-scale industrial rearticulation involving the

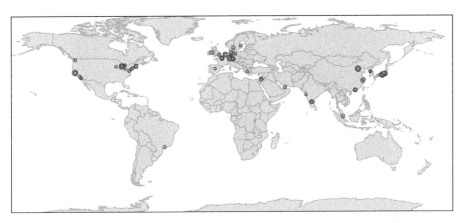

MAP 5. Map showing the geographic distribution of firms in the autonomous vehicle market. Nodes sized by number of corporate transactions (2011–2018). Map elaborated by the author with data compiled by the author and Yuko Aoyama from METI reports and additional sources.

automobile, IT, financial, and other industries.[6] Like ride-hailing platforms, AVs represent a primarily privatized mobility paradigm centered on individual transportation, which has been pushed forward, in the first instance, by IT corporations (including actors like Alphabet (Google) and Uber). Yet, unlike ride-hailing platforms, which presently require human drivers, AVs cater to the long-held dream of eliminating the need for human labor in transportation altogether. Thus, a key aspect of autonomous transportation is the reconfiguration of mobility services (including ride-hailing apps, but also taxis, and even public transportation) from a two-sided market of riders and passengers into one that solely requires passengers. Furthermore, this reconfiguration necessarily impacts another fundamental dimension of the existing mobility market, which is the manufacture and commercialization of automobiles themselves, since it pulls incumbent firms (such as leading automakers, like Toyota, Ford, and GM) into a new arena of competition and technological development. The recent expansion of the autonomous vehicle industry thus reflects a growing interconnection between the automobile, IT, platform, electronics, and other associated industries, resulting in a corporate network with global reach that is nevertheless anchored in highly specialized geographic centers (see Map 5).

The automation of mobility promised by AVs crucially depends on successfully navigating physical space with ever decreasing human input. This in turn relies on the successful integration of an array of mapping,

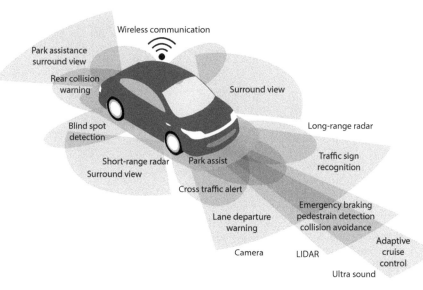

Wireless communication

Park assistance
surround view

Rear collision
warning

Surround view

Blind spot
detection

Long-range radar

Short-range radar
Surround view

Park assist

Traffic sign
recognition

Cross traffic alert

Emergency braking
pedestrain detection
collision avoidance

Lane departure
warning

Adaptive
cruise
control

Camera

LIDAR

Ultra sound

FIGURE 13. Autonomous car remote sensing system. *Source:* Shutterstock.

sensing, processing, and communications technologies (Figure 13). These include GPS, highly precise digital maps, cameras, radar, LIDAR, onboard computers, communications with other vehicles, and connection to digital networks.[7] Different combinations of these and other technologies have given rise to competing approaches to the development of autonomous navigation, as well as a standard taxonomy for driving automation systems. This taxonomy covers six levels of automation, bookended by Level 0, "No Driving Automation," and Level 6 "Full Driving Automation."[8] While the aspiration suggested by Level 6 seems to imply that no humans would intervene in the decision-making process of AVs, this obscures the embeddedness of autonomous driving into broader technological and social systems. Firstly, AVs depend on the coordinated performance of multiple sensing and navigation technologies with near zero error rates—which despite all the hype remains a persistent challenge for large scale deployment and adoption.[9] Secondly, advances in automated navigation notwithstanding, humans intervene—and will likely continue to do so—at various points "in the loop" (and much depends on how said loop is defined) of automated technologies, whether as engineers, operators, supervisors, regulators, or pedestrians. Thus, even if the widespread adoption of AVs remains unrealized, close attention should be paid to this technology's current and future ramifications for cities, regions, governing institutions, and the broader automotive, mobility, and IT industries, all of which are key elements in the reconfiguration of capitalism as this system becomes increasingly digitized.

While the two mobility paradigms embodied by ride hailing and autonomous vehicles are built upon different—though not necessarily mutually exclusive—technical infrastructures, economic models, and social relations, they both require the collection of vast amounts of geospatial data, which is then endowed with economic value and marketized in myriad ways. As it was for many of the other products and services discussed throughout the preceding chapters, the processes of geolocation, valuation, and marketization are crucial for the construction of new mobility industries and the transformation of existing ones—as is the case with the automobile industry. The rest of this chapter explores how ride-hailing platforms and AVs incorporate geospatial data into their operations and how each of these mobility technologies contribute to the construction and expansion of digital capitalism by enacting new combinations of the location, valuation, and marketization processes discussed throughout the book. To contextualize this discussion, in the next section I first address the principal object of these technologies' disruptions: the worldwide system of automobility that took shape throughout the twentieth century.

DISRUPTING THE SYSTEM OF AUTOMOBILITY

Over the past century the automobile has become humanity's principal technology for personal transportation and mobility. The daily routines, consumption habits, and collective imaginaries of billions of people all over the world have been radically altered by the ubiquity of motor vehicles, as have the built landscapes and natural environments at all scales. It is true that other modes and technologies also play important roles in structuring the contemporary global transportation system: most intercontinental trade is conducted via container ships that line the world's oceans: according to research conducted by Statista, "In 2021 the world merchant container ship fleet had a capacity of around 282 million metric tons deadweight. As of January 2020, there were 5,360 container ships in the world's merchant fleet."[10] In many countries, from India and China to Japan, Russia, South Korea, the members of the European Union, and even the United States, trains serve as an important vector for moving people as well as goods on land, both within and between cities. In fact, over the past two decades, the global passenger activity by rail has risen 91 percent, surpassing four trillion passenger-kilometers in 2016—with most of this growth coming from India and China.[11] The scale of air traffic is also enormous, transporting 57.6 million tons of freight in 2019, and reaching a peak of 4.5 billion passengers in the same year (followed by a substantial decrease in 2020 and 2021 due to the COVID-19 pandemic, and a partial recovery in 2022).[12] Yet, as important as these mobility technologies are, their spatial distribution and their integration into the daily lives of people all over the world do not match prominence gained by the automobile.

According to the International Organization of Motor Vehicle Manufacturers (OICA), as of 2015 there were over 1.2 billion motor vehicles in use worldwide (at a time when the global population was 7.405 billion). This is a 30 percent increase in the decade since 2005, when there were only 892 million vehicles worldwide. While the total number of vehicles has continued to grow, it has also expanded relative to global population. During the same decade, the average motorization rate went from 143 to 182 cars per 1,000 inhabitants, an increase of over 27 percent. The geographic distribution of motor vehicles throughout the world shows important variation both across and within regions. According to OICA, in 2015 the region with the most vehicles in use was Asia, Oceania, and the Middle East,[13] with 436.2 million, followed by America (including the NAFTA region, Central America, the Caribbean, and South America) with 413.7 million, Europe with 387.5 million, and Africa with 44.8 million. However, when the motorization rate is considered instead, the most automobile-dense regions are by far Europe and America (including the NAFTA region, Central America, the Caribbean, and South America), with 471 and 418 vehicles per 1,000 inhabitants, respectively. The rest of the world, by contrast, has much lower motorization rates: Asia, Oceania, and the Middle East with 105, and Africa with 42 vehicles per 1,000 inhabitants.

The rise of the automobile as humanity's principal form of transportation did not happen in a vacuum. Rather, it should be understood as both a product of a mature industrial capitalism, as a catalyst of its global expansion, and engine of the ensuing transformation of the planet's environment. In no small measure these dimensions of automobility are reflected in the geographic asymmetries between regions as expressed in the motorization rates cited above. The spread of the automobile and its geographic patterns are manifestations of a historically specific path to industrialization and economic development where the automotive industry acted as a leading edge of capitalist expansion. At least since

Henry Ford's introduction of the Five Dollar Day in 1914,[14] the auto-mobile industry has been instrumental in simultaneously producing and benefitting from a car-centric form of spatial, social, and economic or-ganization; first in advanced capitalist countries, and subsequently in developing ones. Central to this process is the rise of a network of in-dustrial giants that became cornerstones of the national economies of rich countries, harbingers for emerging economies, and in some cases symbols of nationally specific versions of capitalism (such as Fordism or Toyotism). Thus, firms like the Big Three automakers in the United States (Ford, GM, and Chrysler, now Stellantis); Volkswagen, BMW, and Mercedes-Benz in Germany; Renault and Peugeot in France; Fiat in Italy; Toyota, Honda, and Nissan in Japan; and later Hyundai and KIA in South Korea, all contributed to cement not only the establishment of a car-centric form of transportation as the status quo, but the very tra-jectory of twentieth-century capitalism and the geographic transforma-tions that came with it. Among such transformations we can count the construction of highway systems, processes of suburbanization, and the remaking of the built environment across scales—from the street and the neighborhoods to cities, countries, and regions—in the image and needs of the automobile. Altogether, throughout the past century, cars have not only introduced new mobility practices, but in the process, they have also catalyzed profound transformations of the economic, social, built, and natural environments. This sprawling web of changes stem-ming from the massive adoption of cars as a means of transportation can be understood through the idea of an expansive "system of automobil-ity," which has fundamentally reshaped the world around the automo-bile, its needs, and its effects.

Coined by sociologists Mimi Sheller and John Urry, the term *system of automobility* describes how cities, human societies, and indeed an in-creasingly globalized humankind, have been transformed through the intensive adoption and use of automobiles as a central mode of trans-portation.[15] Urry later expanded on this concept and identified six com-ponents that, in combination, "generate and reproduce the specific character of domination" exerted through the system of automobility. These components are: (1) "the quintessential *manufactured object* pro-duced by the leading industrial sectors and iconic firms of 20th-century capitalism"; (2) "the major item of *individual consumption* after hous-ing which provides status to its owner/user through its sign values"; (3) "an extraordinarily powerful *complex* constituted through techni-cal and social interlinkages with other industries"; (4) "the predominant

form of 'quasi-private' *mobility* that subordinates other mobilities";
(5) "the dominant *culture* that sustains major discourses of what constitutes a good life"; and (6) "the single most important cause of *environmental resource-use*."[16]

Considering these six components and their interrelations brings into focus how the system of automobility constitutes much more than a means of transportation: the road and highway networks that crisscross the territories of every country around the globe, roadside attractions and settlements, gas stations, car dealerships, street billboards, auto manufacturing and assembly plants, globally connected supply chains, drive-thru establishments, parking lots, toll booths, emissions testing requirements, speed limits, gas taxes, departments of motor vehicles and licensing processes, stop lights and street signage, auto repair shops, racing circuits, bus lines, and taxi dispatchers. These are but a few of the most visible manifestations of the system of automobility, which is now thoroughly embedded in nearly all facets of human life and our species' imprint on Planet Earth (though with vastly asymmetric sources and impacts).

For over a century the system of automobility has grown to include more and more things, people, places, cultural expressions, resources and, of course, more and more cars—transforming landscapes and enrolling all manner of human (and nonhuman) activities while affecting the environment in potentially irreversible ways. For instance, based on data from the International Energy Agency and the International Council on Clean Transportation, the Our World in Data project reports that in 2018 transport accounted for eight billion tons of CO_2, which make up nearly a quarter (24%) of total carbon emissions from energy. Out of all transport emissions, almost three quarters (74.5%) came from road vehicles, the largest share (45.1%) of which was passenger vehicles (cars, motorcycles, buses, and taxis).[17] Indeed, as suggested by the sixth component in Urry's description of the system of automobility, the ongoing climate emergency cannot be understood without considering the key role of fossil-fuel based automobile transportation brought about by the internal combustion engine as one of the principal sources of carbon emissions.

The specific role of cars in facilitating mobility within, between, and beyond cities varies across geographies, depending on factors like transportation infrastructure and levels of urbanization. However, even in cities with well-functioning public transportation networks (such as metros or streetcars), cars are central to city dwellers' mobility: as private cars, as part of motor vehicle–dependent public transportation (e.g., buses), or as individualized transportation services (e.g., taxis). This is especially true in the United States, where the car became one of the central vectors of geographic change in the second half of the twentieth century. Such changes were concentrated especially in cities, where the automobile was intimately tied with two key spatial transformations: the growth of urban sprawl and urban renewal projects, many of which explicitly focused on building car-centric infrastructure, like highways, at the expense of dense urban neighborhoods, which were often inhabited by poor people and racial minorities. Robert Bullard, whose work helped establish the Environmental Justice movement, summarizes the multiple economic, social, and environmental impacts of the system of automobility on poor people and racial minorities in the United States—many of which are echoed in other parts of the world:

> The disparity of fruits borne by transportation development projects is a grim story of a stolen harvest with disproportionate burdens and costs paid for in diminished life and life opportunities by poor people and people of color. Many federally subsidized transportation construction and infrastructure projects cut wide paths through low-income and people of color neighborhoods. They physically isolate residents from their institutions and businesses, disrupt once-stable communities, displace thriving businesses, contribute to urban sprawl, subsidize infrastructure decline, create traffic gridlock, and subject residents to elevated risks, and explosions from vehicles carrying hazardous chemicals and other dangerous materials. Adding insult to injury, cutbacks in mass transit subsidies have the potential to further isolate the poor in inner-city neighborhoods from areas experiencing job growth—compromising what little they already have. So while some communities receive transportation benefits, others pay the costs. Some communities get roads, while others are stuck with the externalities such as exhaust fumes from other people's cars.[18]

These and other effects of the system of automobility were enabled by the concerted efforts of the state and private capital, and fueled by the development of car culture, the automobile industry as an engine of the US economy, and the path-dependent forces of large infrastructural projects such as the Interstate Highway System, which not only connected the country, but rearranged its geography at all scales, from the urban to the regional. Hence, considering the deep and sprawling roots attaching the system of automobility to humans' way of life in the twenty-first century, any disruption to automobiles as a mobility technology is likely to have outsize effects both for humanity and for the planet. As I argued above, ride-hailing platforms and autonomous vehicles constitute two such disruptions, since they effectively reimagine what automobiles are, how they move through space, and how they integrate into our lives. It is therefore essential to understand how these technologies can change not only car transportation but the system of automobility itself, with potentially transformative implications for the world we have built over the past century. Geospatial technologies are nestled at the very core of these two mobility technologies, and as such they enable not only new ways of understanding and navigating space, but also new ways of bringing space into a capitalist system increasingly defined by its digitization.

The role of geospatial data, media, and technologies in automobility is as old as cars themselves, and in many ways precedes them: objects like road maps, odometers, road signage, and advanced driver-assistance systems have long been crucial for navigation, wayfinding, and other functions of automobile transportation—as they were for modes of transportation that preceded the automobile.[19] The nature and function of geospatial technologies in automobiles has evolved in parallel with the technological development of cars themselves. Over the course of the last half century, cars have gradually incorporated first electrical and electronic systems, and later increasingly sophisticated computers on board. These technological changes have had a broad spectrum of impacts, some of them evident and others less apparent, such as radically altering the operations and capabilities of automobiles, changing the experience of passengers and drivers, and rearranging the configuration of the global supply networks of the auto industry. Innovations like the Electronic Control Unit, and later the wholesale "chipification" of cars created fertile grounds for the embedding of digital geographic information systems in automobile navigation.[20] Today devices like TomTom and Garmin satellite navigators, mapping platforms like Google Maps and Apple Maps, or crowdsourcing applications like Waze are as central to the mobility of billions of drivers as the computerized systems powering and monitoring their cars.

One way of thinking about this shift is that, as cars have become increasingly computerized, they have transformed into mobile spatial media, incorporating navigation, audio, video, geolocation, communication technologies, computerized monitoring, and other digital systems into the primary function of moving through physical space.[21] This transformation of cars not only affects the experience of drivers and passengers but also has the broader effect of drawing cars, and the entire system of automobility, deeper into the logic and workings of the digital economy. In turn, this signifies a fundamental shift in the very structure of global capitalism, particularly since the automobile industry established itself as one of the engines of this system throughout the twentieth century. Therefore, the computerization of cars, and the digitization of many of their functions should be examined with relation to these broader systemic shifts as well as their geographic underpinnings.[22] In the paragraphs below I sketch out some of the implications of two key disruptive forces for the system of automobility, ride-hailing platforms and AVs, with an emphasis on the geospatial dimensions of these mobility technologies.

RIDE-HAILING AND THE REIMAGINING OF SPACE

In many cities around the world taxis have developed for decades into a profitable and powerful industry, even as it is often marked by profound inequalities and an uncertain future. For instance, as reported in the *Guardian*, "according to the New York Taxi Workers Alliance (NYTWA), the average debt owed on medallions by taxi drivers is $600,000."[23] Medallions, which are the physical certificates required to operate New York City's iconic yellow cabs, had long been regarded as prized assets leading to a stable income and a path to social mobility for a largely immigrant workforce. However, the artificial scarcity of the medallions, predatory lending practices around their financing, and speculation surrounding their price fluctuations created an inherently unstable market. Drivers were in a particularly precarious position since they were saddled with ever-expanding debt to cover the rising price of medallions. According to the *Financial Times*, a bubble began to form in the taxi medallion market in the late 2000s, only to crash a few years later with disastrous consequences: "Medallions in the early 2000s had changed hands for $300k. A decade later the price had tripled. Yet, by the end of the 2010s, the taxi license was changing hands for just over $100k leaving many individual owners well under water."[24]

The downward turn in this market came around 2014 after the entrance of TNCs like Uber and Lyft into New York City, foreshowing a radical change in the transportation landscape. High levels of debt coupled with plummeting medallion values and new competition from TNCs left thousands of drivers buried under mountains of debt with little to no possibility of repaying. After years of protests, driver suicides, and even a hunger strike, in 2021 a deal was reached between NYC's largest driver organization, the New York Taxi Workers Alliance, and the largest loan holder, Marblegate Asset Management. This deal reduces medallion debt to a cap of $170k per driver, while the city will pay the lender $30,000 for every loan and also act as guarantor of the debt. The total amount expected to be paid by the city is estimated at $65 million.[25]

The taxi medallion crisis in New York City illustrates some key dynamics in the taxi industry that in turn can help us contextualize and understand the disruptions unleashed by new mobility technologies. First, the asymmetric power relations between municipalities, private financiers, and taxi drivers have in this case been characterized by exploitation of vulnerable populations through the manipulation of the price of a key asset: the taxi medallion. Even as a deal was reached to improve the debt payment conditions for drivers, the very solution to the medallion crisis is once again seen as a profit opportunity from new financial products. As reported by the *Financial Times*, "After the restructuring deal, more mainstream Wall Street institutions are beginning to see a way that they could play in the taxi ecosystem too. One person involved in the taxi medallion market speculated that following the restructuring, medallion debt would become standardized enough to be packaged in a securitized product. This person said the likes of Credit Suisse and Goldman Sachs have expressed preliminary interest in the medallions."[26]

Another key lesson from the medallion crisis is how it highlights the increased precarity of drivers in a profession that was once seen as an avenue for upward social mobility in cities in the United States and beyond, particularly for immigrants. Precisely due to the organizational, financial, and regulatory arrangements that created and inflated the medallion market, many drivers became trapped in unsustainable situations from which it became very difficult to escape, something which only found a remedy after concerted collective action and intervention from the state. A third key point is how the medallion crisis reinforces both the imperiled status and the reputational stigma of the taxi industry itself. While medallions were at the center of unsavory practices of many lenders and the prolonged lack of response from the municipality, this cannot be seen as an isolated incident. The taxi industry in NYC and in many cities around the world has periodically been mired in scandals that involve not only poor organizational, regulatory, and financing practices, but also widespread notions of insecurity for passengers as well as drivers. Together these three factors set the stage for understanding why the entrance of TNCs like Uber and Lyft had such an immediately transformative effect in city after city. A cornerstone of the strategy of TNCs was positioning themselves in contrast to the taxi establishment, playing on the public's perception of the three factors outlined above and promising a radically different mobility experience: safer, more modern, technology centered, and free from the constraints and stigmas plaguing the taxi industry.

In Latin American cities, for instance, taxis continue to be prevalent, but they are frequently viewed as risky for passengers. In this context "radio taxis" have become established as an alternative, especially for the middle classes. For a price differential, radio taxis provide fleets of cabs that can be summoned by passengers via phone calls, rather than hailed from the sidewalk. Radio taxis can be seen as a precedent to the services provided by TNCs: passengers can establish prior contact with the cab, and by using the dispatcher as an intermediary, a record is created that can serve to build trust and enhance the perception of safety for passengers. In the context of radio taxis, location becomes an invaluable asset, since it is a key feature to establish trust: passengers communicate directly with a trusted location, the taxi hub, which then dispatches a verified vehicle to the agreed-upon pickup site.

The trusted location and communication offered by radio taxis constitute important elements for the valuation of this mode of transportation by passengers, and therefore to the construction of a parallel taxi market. However, even though the information exchanged between passengers and radio taxis increases the level of trust and transparency in comparison to regular taxis, there remain two key spatial gaps: (1) the path between the taxi's point of origin when dispatched and the pickup site, and (2) the path between the pickup site and the passenger's destination. It is in those very spatial gaps that new kinds of location data, new valuation models, and new markets have emerged in the form of TNCs. As I will explore later in this chapter, TNCs build on the structure of intermediation provided by radio taxis and add digitally mediated features such as real-time tracking of navigation through space and access to information about the car, driver, and route during each trip within their digital mobile applications. Together these factors place TNCs at an advantage over radio taxis (not to mention regular taxis), particularly because ride-hailing applications allow passengers to reconceptualize their navigation through space as knowable, organized, and predictable. This in turn leads to the new integration of GPS-enabled location and tracking data with new forms of valuation, payment, and labor practices overseen by the TNCs, and in turn, the creation of new transportation markets that are disrupting not only urban mobility but the very social fabric of cities the world over.[27]

An illustrative example of the transitions experienced by the taxi industry, and urban mobility more generally, throughout the introduction of TNCs is the case of Buenos Aires, documented by economist and anthropologist Juan del Nido. The regulation of taxis in Buenos Aires has been a contentious process at least since taxi licenses were required after the murder of a passenger in 2001.[28] Then, in 2013 "a pedestrian's death in an accident involving a taxi driver in an alleged hypoglycemic shock presented the industry with the question of how to minimize the likelihood of such a thing happening again."[29] This case catalyzed a process of increased inspections, driver certifications, and other management and regulatory tools that led to the transition of the taxi industry in Buenos Aires throughout 2015 and 2016.[30] Such transitions were all marked by contentious politics involving the city government, the taxi drivers' union—whose leader was influential in the national government—and even the country's presidency. It was in this context that Uber arrived in Buenos Aires in April 2016. As del Nido describes in rich ethnographic detail, Uber symbolized the global aspirations of the *porteño* middle class, even as it encapsulated a fundamental irony: the very service that was promised as an alternative to the reputational stigma of the taxi industry was itself legally challenged and immediately mired in continuous court battles. Setting off litigation upon arrival was core to Uber's "disruptive" strategy, described by legal scholars Ruth Berins Collier, Veena Dubal, and Chrisopher Carter as a "brash 'act first, apologize later' entrance into urban markets." Collier, Dubal, and Carter argue that this strategy has at its core a "model of disrupted regulation, which has two phases":

> In the first, an existing regulatory regime, in this case for taxis, was not deregulated but disregarded by the challenger, Uber, who flouted entry and price controls, often triggering cease and desist orders from city regulators. A subsequent phase involves regulation and has occurred at both city and state levels—in legislative and sometimes regulatory bodies—and also in judicial venues. It conforms to an elite-dominated model of contending incumbent vs. challenger interests, in which the latter has largely prevailed. In this model of challenger capture, Uber has been able to defend its core interests of low prices, high driver supply (with no labor regulation), and consumer trust. While Uber initially rejected all regulation, it has most vigorously opposed those central to its business model of low-cost service with dynamic pricing, frictionless entry of drivers, and no vehicle caps.[31]

After Uber began providing service in Buenos Aires the battles in Argentine courts played for years and were closely tied to the political winds that affected transitions in the national government (Figure 14). The disproportionate political salience of Uber can be in part attributed

FIGURE 14. Protests against Uber in Buenos Aires. Photo by Yair Cohen, https://creative commons.org/licenses/by/2.0/#.

to the fact that the company and the service it provided were symbolic of a particular idea of private enterprise which was a point of contention between rival political parties who alternated in power. As of this writing, Uber continues to be operational in Buenos Aires, having been declared legal in September 2020.

Beyond constituting a new form of mobility, del Nido contends that Uber presented an epistemic break not only for passengers, but for broader swathes of *porteño* society. This epistemic break revolutionized how the people of Buenos Aires—and particularly a middle class with global aspirations—came to understand their possibilities of movement through, and habitation of, the city. He calls this a "Copernican phantasmagoria," where "the middle class understood Uber relations as an "ordered, orderly order" made of propositions in a grammar of efficiency, supply, and demand that revolutionized how movement could be inhabited and known from within in Buenos Aires." This phantasmagoria was not only a rhetorical construction, but a technological one, since it was "produced by the company's interface for each and every one of its users, these propositions were unverifiable and could not be debated: one accepted or declined them."[32] A defining characteristic of this phantasmagoria was the illusion that passengers were able to "see everything" surrounding their trip (e.g., driver names and photographs, car plates, car types, ride prices, etc.).

Extending del Nido's argument, I posit that the ordered illusion of Uber's "Copernican phantasmagoria" constitutes a form of spatially rooted knowledge anchored in the cartographic rationality and mediation provided by the application's mobile interface. In this way, the city, and space itself, becomes apparently knowable, manageable, and navigable through an interactive, continuously updated, and mobile map that conditions one's movements in space before, during, and after the completion of a trip (Figure 15). This spatial knowledge is in turn tied to more fundamental ideas about freedom of movement, individuated mobility, and one's place in a socio-spatial order. Yet this new spatial knowledge offered by Uber rested on a fundamental contradiction: even while passengers seemed to embrace it and justified it in terms that contrasted with the perceived obscurity and untrustworthiness of the taxi industry, neither the very logic behind this new ordering nor the operations that made it possible were themselves transparent or knowable. In fact, as has been argued by researchers, activists, and journalists alike, Uber and many other tech companies thrive on information asymmetries, even as they foreground a rhetoric of transparency.[33]

As del Nido discusses, while some of his informants were able to speak very eloquently about base rates and multipliers for the price of Uber rides, they were unable to articulate where those multipliers actually came from or describe specifically how they worked. This is not necessarily their fault, as Uber's algorithms, like those of many other firms operating in the digital economy, are secretive by design. Yet what stands out is the illusion of certainty produced by Uber's application not only in the navigation through space but in passengers' knowledge of how the world around them is ordered and organized.

This illusion of certainty is present in another aspect of Uber's "Copernican phantasmagoria" and the spaces it creates: the optimization of routes during rides, and particularly how this optimization often diverged from both the drivers' and passengers' knowledge and experience of space. One of the instances recalled by del Nido is the experience of Valentina, a passenger who frequently took Uber to and from her work.[34] On Uber's application map the route would consistently direct the driver to take a particular street, Marcelo T. de Alvear, known to *porteños* as narrow and busy at all hours with traffic from businesses, tourists, and offices. During Valentina's Uber trips the drivers would routinely avoid this street, even though it was often indicated as part of the application's automatically calculated route. Upon avoidance, the initial route was immediately recalculated on Uber's map. In each case neither the drivers nor Valentina paid too much attention to this change, since as locals they all knew to avoid that street. At the end of the trip, when prompted to review her passenger experience and rate the driver, Valentina would not give a lower rating as a result of the street avoidance— even if in this case she technically had grounds to do so, according to the "optimized" route initially indicated by Uber's algorithm.

This tacitly agreed-upon systematic avoidance of Marcelo T. de Alvear Street illustrates some of the fundamental reconfigurations of spatial and social relations that result not only from the use of Uber's application, but more specifically from its reliance on geospatial technologies to generate an algorithmically ordered experience of space. First, as del Nido points out, the route initially calculated by Uber represented the "ordered, orderly transaction produced specifically [for Valentina]," which in turn is part of a "seemingly objective grid" through which her passenger experience is organized, communicated, and evaluated.[35] Yet this process of representation, interaction, and knowledge production is in many ways an illusion (a phantasmagoria) because in the end all that matters is not whether the passenger experience adhered to the parameters set by the aforementioned "grid" but Valentina's own personal preferences. In other words, even as Uber produced a representation of seemingly knowable and organized space that was then used to evaluate the experience of navigating from point A to point B, said representation was secondary. Underlying this experience of mobility was the process of individuation and privatization that allowed each actor to issue a judgment entirely independent from the "grid" against which the experience is purportedly evaluated.

This tension between individuated mobility experiences held together by each ride can be expressed in the ratings issued of both passengers and drivers after a trip, a practice common across TNCs and ride-hailing applications around the world. While appearing to provide an objective evaluation of "quality," the ratings also collapse an uncountably complex set of interactions in ways that hinge on subjective preferences, arbitrary or reasonable as they may be. Both passenger and driver have the ability to deduct stars from a rating for any reason and without explanation. In this way, an unwelcome remark, an impolite expression, or an instance of careless driving can represent the same rating deduction as the wise avoidance of a known narrow street or a passenger's idiosyncratic dislike of the driver's physical appearance. This privatization of experience mediated by metrics and backed by the "grid" and the "map," which makes it appear authoritative and objective, is one of the hallmarks of

FIGURE 15. Ride-hailing mobile application interface. *Source:* Shutterstock.

the platform economy and is particularly salient in the form of mobility mediated by ride-hailing applications. Related to this, a second point raised by Valentina's anecdote is the *blackboxing* of power relations that pervades ride-hailing transactions, and by consequence, the markets built around such transactions. As suggested above, del Nido's interlocutors were not able to articulate where price multipliers came from, even as they fundamentally believed that they took part in a mobility experience structured by a transparent market where all actors are empowered by information and knowledge. However, as both the unknowable reasons behind price multipliers and the route calculation algorithms show, platforms like Uber provide a sense of agency, knowledge, and control to all parties involved in the mobility experience. Yet, at the same time, such platforms leverage the very tools through which they create this sense of openness and agency to generate economic value, accumulate power, centralize data, and obscure decision-making through the creation of information asymmetries—all of which in turn lead to monetization and growth, if rarely profitability.

In the case of ratings, even as their purpose is to uphold certain standards of interaction in the ride-hailing process, they do not account for the complex power geometries that determine each moment in the interaction between passenger and driver. For instance, while a passenger may be in a more vulnerable position when riding in a stranger's vehicle, a driver's ability to earn a living is continuously on the line with every rating, leading to a situation of ongoing precarity. By offering a process to screen drivers and passengers, provide real-time monitoring of routes as the car moves through space, and ensure certain behaviors, Uber has presented itself as an alternative to the taxi industry—a strategy it deployed successfully in Buenos Aires and replicated, with local variations, in over ten thousand cities around the world.[36] An important part of this promise of alternative mobility relied on the provision of transparency through the enactment of transactions knowable to all parties. Regardless of the degree to which these goals were actually accomplished, what was of greater consequence for Uber's successful deployment in Buenos Aires—as in other cities around the world—was the collective investment in the legibility and knowability of the "Copernican phantasmagoria" described by del Nido, particularly by actors in society with global aspirations. Consistent with ascendancy of a "technological ideology" that is particularly salient in urban life, such aspirations were articulated through a belief in the capacity of new digital technologies to solve complex social problems, such as the decades-old conflicts around the taxi industry in Buenos Aires.[37]

Examined through the lens of the LVM framework, the operations of Uber, and to a certain extent those of other TNCs throughout the world, provide a window into new forms of integration between geospatial data, the creation of new spatial forms of knowledge, and the construction of new mobility markets. First, the real-time location collected by smartphones is integrated into Uber's interface to fill the spatial gaps left by previous mobility services, like radio taxis. This in turn constitutes a source of value, since it makes space and movement more knowable for passengers and drivers, even as it renders other aspects of mobility unknowable. By the same token, Uber, as the intermediary platform, centralizes knowledge about all parties, trajectories, and transactions involved in each ride, which puts it in a favorable position to decide the generation and allocation of economic value. These processes of location and valuation, in turn, act in concert with Uber's own corporate strategy of disrupting urban transportation, contributing to laying the foundation for new mobility markets built on the intermediary capacities of the ride-hailing application and its associated spatial transformations.

Accordingly, even in the face of staunch opposition from taxi groups, and often the local government, the entire mobility landscape in Buenos Aires began to pivot toward the services offered by Uber and other TNCs. While radio taxis still exist in cities from Buenos Aires to Mexico City, the communication and trust-building services they provided was ultimately afflicted by spatial gaps that newcomers were able to exploit. When TNCs began to move into Latin American cities, firms like Uber (USA), Cabify (Spain), 99 (Brazil), or DiDi Chuxing (China) provided services that could essentially be seen as an upgraded version of radio taxis, albeit with key differences. For one, TNCs addressed many of the safety and security issues of regular taxis. However, instead of relying on the telephone like radio taxis, ride-hailing apps created a legible spatial experience for drivers and passengers. This system relied on leveraging the capabilities of mobile digital applications, starting with GPS geolocation available in smartphones. In this way, ride-hailing apps filled the spatial gaps left by radio taxis by integrating location tracking with various layers of verification, trust-building, and communication (among others), offering passengers and drivers the possibility of a complete view of each ride, and even the possibility of monitoring the conditions of the market for rides in real time. Hence, the kind of mobility experiences that TNCs introduced, and how they were able to outcompete both traditional and radio taxis cannot be separated from the reconfiguration of spatial, social, and informational relations enacted by ride-hailing apps. This reconfiguration is at its core made possible through the use of geospatial technologies not only as a way of navigating space, but also as a way of organizing it into an experience that seems knowable, legible, and transparent to all parties—even as this heavily relies on the use of obfuscation and blackboxing leveraged by TNCs and their platforms in their roles as intermediaries.

Such information asymmetries characterizing the mobility transactions behind each ride also shape the spatial strategies that Uber and other TNCs deploy to allocate "assets" in space. As Wells, Attoh, and Cullen argue, Uber relies on a "just-in-place" strategy to ensure the optimal distribution of passengers and riders in a way that maximizes efficiency and earnings for the TNC: "the big innovation of the Uber platform is the creation of a "just-in-place" worker. Akin to those materials for assembly lines that arrived just-in-time for production, so too do drivers end up in just the right place for Uber's services to be offered.

The Uber platform relies on algorithms not to schedule its drivers as much as to place its workers where it wants them across the city."[38] One of the means through which Uber and other TNCs enact this spatial distribution is through a technique called geo-fencing, through which they "keep track of and sort data, which in this case is spatial and temporal information about how ride requests should be dispatched."[39] However, this strategy of spatial allocation has the unintended effect of creating conditions where workers can begin to circumvent Uber's own underlying logic of labor fragmentation precisely because it creates pockets of spatial—and sometimes, social—proximity. When drivers are waiting in the same area for long periods of time, day after day, like outside of airport terminals, they may begin to strike conversations, get to know each other, and develop social bonds. Thus, as Wells, Attoh, and Cullen demonstrate through their examination of Uber protests in Washington, DC, the very spaces created by Uber's spatial strategy of "just-in-place" labor (such as airport parking lots) can also become the very sites where solidarity and worker organization can begin to gain momentum.

Altogether, the widespread popularity of TNCs and their ride-hailing platforms, along with the emergence of autonomous vehicles together constitute important disruptions into existing systems of transportation and mobility, particularly in cities. Both of these depend on new forms of data collection via diverse arrays of sensors from cameras and LIDAR to GPS. Such data streams are in turn enrolled in new mechanisms of location, valuation, and marketization through their integration with new forms of spatial knowledge, corporate strategies, and socio-spatial relations. Yet neither of these constitute a serious challenge to the system of automobility and thus cannot be considered vectors for transition into fundamentally different mobility regimes. After all, as Mimi Sheller reminds us, mobility regimes must be understood as part of much broader contexts:

> Mobility regimes are embedded in entrenched repertoires for political interaction, transport planning and urban governance. And urban governance itself does not occur in a vacuum but relies on the extended "operational landscapes" of resource extraction, connected to energy generation, food systems, water systems, mining, oil extraction, military power, etc. Transitions in mobility systems depend, therefore, not just on individual choices, technological transformations, or even "disruption," but also on transitions in entire kinopolitical cultures.[40]

Taking Sheller's insights seriously requires reckoning with the fact that, while geospatial data, media, and technologies are central to analyzing the maintenance of existing mobility regimes and the emergence of new ones, they are also insufficient to properly understand the broader contexts of those regimes. To do so, it is necessary to connect the discussions of the spatial architecture of capitalism, and its underlying processes of location, valuation, and marketization, with discussions about the concrete places, spaces, and territories where such processes unfold. This involves leveraging the very geography upon which digital capitalism relies toward critical analysis and various forms of intervention, from regulation to political organizing. The concluding chapter of this book offers potential approaches to achieve this goal.

Conclusion

The Spatial Architecture of Digital
Capitalism and the Power of Place

Geospatial data, media, and technologies are not inherently digital, and their existence far precedes today's digital turn. Yet, just as artifacts like the surveyor's chain and the cadastral map transformed notions of property and even enabled the expansion of empires and modern regimes of private property, the incorporation of geospatial technologies into digital capitalism both reveals its underlying spatial architecture and enabling new transformations. In this book I have given an account of how maps and the broader category of geospatial data play fundamental roles in structuring digital capitalism. As the examples discussed in the previous chapters illustrate (from Google Maps and IP geolocation to ride-hailing platforms and remote sensing satellite data), the various forms of integration of geospatial data in the digital economy are also key to understanding the ongoing transformations of our economic landscape and the politics and policies surrounding such processes.

At the core of the arguments developed throughout this book is the notion that the digital economy is not (and has never been) abstracted from place, space, or territory, but rather is thoroughly entangled with geography. However, the appeal of what Vincent Mosco has termed "the digital sublime" has for decades fueled the construction of narratives that disassociated the workings of digital capitalism from its geographic underpinnings.[1] While these narratives, such as the notion of "cyberspace" indeed obscured many of the geographic dimensions of the digital, it must also be acknowledged that they were able to succeed in part because the affordances of digital environments did not leverage geography in explicit ways until relatively recently. Once geospatial tools like IP and GPS geolocation were more intentionally brought into the digital economy in the early 2000s, new products, services, and markets emerged that directly responded to geographic variation and context. By the same token, this shift also highlighted the spatial underpinnings that have shaped the digital economy from its very inception. It is precisely because geospatial technologies enable the linkage of information "flows" with concrete geographies that we must understand the evolution of these technologies, their variations, and interconnections with broader processes at the core of digital capitalism. To this end I have proposed examining digital capitalism through the conceptual triad of location, valuation, and marketization (LVM). As I have shown by examining various emerging industries and technological developments in the preceding chapters, the LVM conceptual triad can help elucidate both how the digital is fundamentally (even if not always visibly) geographic, and how the geographic is central for the logic and workings of digital capitalism. The connecting thread in this case is revealed by paying particular attention to those elements that provide the explicit links between information flows and concrete geographies: geospatial data, media, and technologies.

Acknowledging the fundamental role that geospatial data, media, and technologies have played in building a particular version of digital capitalism is therefore also essential for understanding how it works and how its benefits and harms are distributed. Furthermore, it is by clearly establishing the geographic underpinnings and the geospatial workings of digital capitalism that we can also leverage the power of place to reframe it, rethink it, and transform it in ways that better suit what we might like to see across different contexts, communities, and constituencies. For instance, since so many systems of surveillance, control, domination, discrimination, and exclusion operate in thoroughly digitized ways, the task of building a more just world forces us to reckon with the specific mechanisms by which digital technologies are implemented to those ends.[2]

This task requires identifying specific avenues to assert collective means of deliberation, regulation, and control over the processes and actors who have accumulated and exerted digital power. To an important extent, this power accumulation has grown unchecked because we often miss how it is intertwined in immediate ways with the places, locations, and spaces of our everyday lives as well as those spaces where we can exert collective power. Recent turns toward anti-monopoly regulation in digital markets in the United States, important (if insufficient) advances toward the protection of personal data in jurisdictions like the European Union and California (which rely in significant ways upon the geographic logic of establishing the location of data as a means for regulating it), are examples of the power that place-based institutions, governance structures, and regulatory frameworks can exert over an environment that too often has been framed as consisting of ethereal and placeless flows of data, content, and capital.

As I have argued throughout the book, three key processes to understand how geospatial data contribute to structure and operate the digital economy are (1) location, (2) valuation, and (3) marketization. Attending to these processes can help us understand the workings of digital capitalism while also helping us identify openings to reimagine and transform it. To this end, it can be helpful, as I have suggested in previous chapters, to start from the infrastructural dimension. A geographical understanding of digital capitalism should therefore build on the findings, insights, and arguments of scholars, journalists, artists, and activists who have thoroughly documented how, for instance, the internet is not something that exists abstracted from the physical environment or even the political organization of the world, but a construct that is as much material, political, and economic as it is cultural, symbolic, and rhetorical.[3] Some of this information network's geographic manifestations have become better known over time, such as undersea cables, data centers, and cell towers. However, the spatial roots of the internet, and the digital capitalism it underpins, can be traced to its inception. Beyond its infrastructural dimensions, it is the very logic of spatial constitution and organization underlying the internet that must be brought to the fore, especially since many of its long-term reverberations continue to shape digital capitalism in the second decade of the twenty-first century.

From its earliest iterations, the projects that would become the internet were created to overcome specific spatial challenges that arose during the Cold War. In particular, ARPANET and other precursors of the internet were conceptualized and designed to circumvent the communication breakdown that could happen in the case of a nuclear attack on US territory. This marked the emergence of a communications system irremediably entangled in Cold War geopolitics, the territorial configuration of the United States, and the production of knowledge and allocation of funds at the nexus of the military industrial complex, the scientific establishment, and higher education institutions. Thus, from the beginning, the infrastructure of the internet followed a specific spatial, institutional, and economic logic closely tied to key factors like defense spending and the pivotal role of universities and research labs in developing the protocols, tools, and systems that underpinned the creation of today's digital infrastructure.

Over the decades, this infrastructure has expanded dramatically through the connection of new networks from all over the world—each with their own logics and histories—as well as its diversification, growth, and increasing influence of a robust private sector.[4] This new environment has in turn brought about important changes to the informational, physical, and social landscape of the internet. Today the global corporations that command large shares of our digital economy also control vast infrastructure networks that include data centers, undersea cables, large real estate holdings, fulfillment centers and many other elements. This is a direct result of the pervasive digitization of products, services, activities, and interactions that make up a large part of our lives. As such, it is necessary to explain how and why digital capitalism is transforming the places and spaces where we live, work, vote, and play. By the same token, given that these spatial transformations entail the concentration of power, resources, and the unequal distribution of benefits and harms, it is also necessary to mobilize such explanations to demand transparency, accountability, and change. To do so, rather than rely on the disaggregated and ethereal imagery promoted by notions like "the cloud," we must turn to the structures and institutions that leverage the power of place: from local, state, and national governments to regulatory agencies and community organizations to universities and civil society groups.

Leveraging such place-based structures and institutions can pave the way for conversations, investigations, and regulations addressing fundamental questions at the heart of digital capitalism and its influence in our lives. This may entail addressing questions such as what are the political, administrative, and geopolitical considerations involved in the siting of a data center within the jurisdiction of a particular country? What are the social and environmental externalities of bitcoin mining? Should governments demand the physical presence of internet companies within a country's borders as a requirement for doing business in the digital economy? How should digital commerce, social networking, and other digital activities be taxed, and who should be the main beneficiary? Are gig workers fundamentally different from employees in any other industry, or are they entitled to the same employment provisions and protections? These are all questions that emerge from the workings of digital capitalism, but whose answers are not fundamentally found in the digital sphere. Indeed, reasserting the power of place is a way to bring the digital back into the arena of politics and the everyday, a way to pierce through the veil of aspatial obfuscation that for decades has shrouded the mythical "flows of information" that make up other constructions like "our global economy," the "information superhighway," and "our digital future." Elsewhere, in my work with Jovanna Rosen, we have proposed that a way of emplacing digital capitalism is by examining what we call "the digital growth machine." This idea refers to how the growth coalitions of political and economic interests at the core of cities mutate when they incorporate the tools, logics, and actors from the information technology industry.[5] We posit that the digital growth machine creates new avenues for capital accumulation that transform not only cities but also their hinterlands. As a result, urban and nonurban places take on the image and priorities of powerful actors in digital capitalism, often with results that accentuate inequities within and between those places. More broadly, the ascendance of the digital growth machine changes how cities and regions operate, modifies the logics and systems that govern them, and sets the terms for who can benefit and who is left out in the space-economy built around digital capitalism. Thus, focusing on the dynamics of the digital growth machine is one way of bringing the digital down to Earth and studying its concrete consequences for concrete places and people. Table 1 summarizes the key dynamics of the digital growth machine and the avenues of capital accumulation it promotes.

My hope is that the ideas developed in this book may prompt scholarship and conversations toward other ways of emplacing digital capitalism and reshaping it in a fairer and more equitable direction. Throughout the preceding chapters I have argued that geospatial data, media, and technologies represent a powerful analytic to reframe the dominant narratives of digital capitalism—and bring the digital back into the geographic. This is because such elements undergird the myriad mechanisms by which digital capitalism has been able to exploit the power of space, place, and locality. The development and digitization of geospatial technologies has led to new products and services that cater to the specific preferences of people according to their location, track their movements, help them navigate from place to place, and watch their lives from above. There may be useful as well as nefarious applications of many of these products and services, and there are specific regulatory questions that should be asked for each case. More fundamentally the underlying spatializing logic embedded in geospatial technologies can be turned on its

TABLE I   CORE COMPONENTS OF THE DIGITAL GROWTH MACHINE

Forces underlying the ideology of technology	Increased reliance on technology-sector attraction as an engine for economic growth.
	Technological solutionism that reframes urban problems as technological problems requiring technological solutions.
New capital accumulation pathways	Extends long-standing land development and industrial attraction strategies to promote urban growth and to increase exchange values.
	Develops new possibilities for capturing land-related profit beyond traditional land development and intensification strategies.
	Supports new opportunities for intermediaries to emerge and profit by monetizing different aspects of urban life.
	Creates new digital renderings of the city that affect land- and asset-related value.

NOTE: Adapted from content in Jovanna Rosen and Luis F. Alvarez León, "The Digital Growth Machine: Urban Change and the Ideology of Technology," *Annals of the American Association of Geographers* 112, no. 8 (2022): 2248–65.

head to analyze, critique, and transform digital capitalism, which it helps to maintain.

In its contemporary incarnation, digital capitalism can be experienced in myriad location-aware ways that mesh digital information with spatial context. This is possible due to technological breakthroughs like the unscrambling of GPS signals in May of 2000 to increase the public's access to more precise location, and the development of IP geolocation technologies that tie IP addresses of specific computers to physical locations. Before these innovations, digital environments were less immersive, more abstracted, and seemingly aspatial. Today by contrast, users' informational experience is often mediated by search engines and other tools that sort results in part by location and proximity. That many such tools are increasingly powered by artificial intelligence systems brings yet another layer of complexity into the spatialities of digital capitalism. Yet, even as it would seem that increased automation makes the digital even more abstract and removed from human agency, AI is inevitably emplaced, as it enrolls specific people, firms, resources, locations, and governments into its production, testing, deployment, and consumption.[6] This brings geography into the digital economy in much more visible and experiential terms than ever before. A consequence of this is the creation of new spatial dynamics while simultaneously bringing to the surface some of the spatial logics that were deeply embedded in the internet conceptualization and execution from its very beginnings.

Precisely because they can now be more explicitly tied to location in context, geospatial technologies have become a key mechanism through which data can be imbued with economic value, turned into commodities and assets, and used to build markets of various kinds. That means that new commodities, new assets, and new services can explicitly incorporate space and derive value from it: from targeted advertising in platforms like Facebook and Google to navigation tools, sorted search engine results, and geoblocked media libraries in subscription services like Netflix. In all these cases information is pinned to geographic location, and tailored to specific places, which allows for new economic relations such as the customization of digital content. However, these very tools also allow for power relations of different kinds such as digital surveillance, discrimination, harassment, censorship, and cyberattacks.

Once information is linked to specific locations, as I have shown throughout the book, we should examine how it is imbued with economic value in highly contextualized ways and how it circulates in specific markets. Those markets never exist in a vacuum, but they are constructed by and through geographic factors and shaped by institutions rooted in particular places. Furthermore, digital markets must also contend with physical, material, economic, and political constraints that vary across times and places—from the local and regional to the national and global. In sum, digital information markets, and digital capitalism itself are thoroughly constituted through the building blocks of geography such as location, place, space, territory, and scale. This means, for instance, that the version of digital capitalism unfolding at a neighborhood scale in the Boston metropolitan area may have some systematic connections and similarities to that which operates across New England, or even throughout the United States. However, each of these versions is also shaped by myriad localized combinations of contextual factors that result in a landscape of variegated digital capitalism: a vast collection of digitally mediated economic formations that, though different in many respects, are all joined at the root through the fundamental workings of capitalism. In other words, while digital capitalism may at times act as a force for homogenization, it nonetheless derives much of its value from spatial heterogeneity.

What are the stakes of reframing digital capitalism as a thoroughly geographic construction, starting with a focus on the crucial function played by geospatial technologies and their role in the processes of location, valuation, and marketization? First, our lives are already shaped by geography because, even when we are mobile or wirelessly connected with others at a distance, we still live in concrete locales, and we experience the world according to the places where we spend time, where we have political rights, where we can access services and participate in social processes of different kinds. All of this is mediated, though not determined, by factors like location, place, space, territory, and scale. And even though digital technologies often obscure spatial relations, such factors also mediate our interactions and experiences in the digital sphere.

While for a long time the digital has been discursively separated from geography and location through a pervasively aspatial imaginary, there are some signs that this idea is waning, especially as events like the COVID-19 pandemic have brought to the fore how our lives are dependent on location and the digital is not a substitute for "being there," but rather interacts with our spatial context in complex and ever-changing ways. Therefore, if we reframe our understanding of digital capitalism to reflect how it is influenced by geography, we will be better equipped to answer pressing questions for our shared future, such as: Under what conditions is "the economy" built, and how can we reimagine who the winners and losers are? Who is included in an increasingly digitized future, and who is excluded? Which countries, regions, corporations, and constituencies are favored in present and future iterations of digital capitalism? Which groups of people are systematically excluded? And how is this exclusion mediated through digital technologies?

Often our experience of the digital can feel seamless, but as I have shown throughout the book there are layers of labor, infrastructure, economic transformations, social rearticulations, and discursive formations that make it appear so. Paradoxically, all these layers have deep geographic expressions and histories. In fact, at the core of today's digital capitalism is a set of processes of location, valuation, and marketization that simultaneously rely on the richness of geographic context, while often obscuring it from view. Asserting the power of place to reframe digital capitalism in geographic terms is a way to reclaim the digital and steer it toward alternative logics, values, and priorities that better respond to how we live our daily lives, how we may want to organize our social and political world, and how we want to confront some of the most urgent collective challenges that lie ahead.

Notes

CHAPTER 1. INTRODUCTION

1. At the time of writing (May 2023) only one Blockbuster location remains, in Bend, Oregon, down from a peak of 9,094 stores in 2004.

2. Lisa Richwine and Dawn Chmielewski, "Hollywood Writers Strike over Pay in Streaming TV 'Gig Economy,'" Reuters, May 2, 2023, https://www .reuters.com/lifestyle/hollywood-writers-studios-stage-last-minute-talks-strike -deadline-looms-2023-05-01/.

3. Josh Rottenberg, "Hollywood Actors Join WGA in Historic Double Strike. 'This Is All of Our Fight,'" *Los Angeles Times*, July 14, 2023, https://www .latimes.com/entertainment-arts/business/story/2023-07-14/actors-strike-sag -aftra-joins-writers-guild-picket-lines.

4. Susanne Elizabeth Freidberg, *Fresh: A Perishable History* (Cambridge, MA: Harvard University Press, 2009).

5. "The closest meaning of *dabbawala* in English would be "lunch box delivery man." Abhishek Chakraborty and Akshay Narayan Hargude, "Dabbawala: Introducing Technology to the Dabbawalas of Mumbai," in *Proceedings of the 17th International Conference on Human-Computer Interaction with Mobile Devices and Services Adjunct* (MobileHCI '15: 17th International Conference on Human-Computer Interaction with Mobile Devices and Services, Copenhagen Denmark: ACM, 2015), 661, https://doi.org/10.1145/2786567.2793685. Since the word *tiffin* is another way of referring to lunchbox, *dabbawala* are also often called *tiffinwalahs or tiffinwalas*. Gauri Sanjeev Pathak, "Delivering the Nation: The Dabbawalas of Mumbai," *South Asia: Journal of South Asian Studies* 33, no. 2 (2010): 236, https://doi.org/10.1080/00856401.2010.493280.

6. Pathak, "Delivering the Nation," 236.

7. As told to Perry Garfinkel, "Delivering Lunch in Mumbai, across Generations," *New York Times*, February 2, 2017, sec. Job Market, https://www.nytimes.com/2017/02/02/jobs/dabbawalas-india-lunch.html; "The Cult of the Dabbawala," *Economist*, July 10, 2008, https://www.economist.com/business/2008/07/10/the-cult-of-the-dabbawala.

8. Silvia Federici, *Wages against Housework* (Bristol, UK: Falling Wall Press, 1975); Silvia Federici, *Caliban and the Witch*, Second, revised edition (Brooklyn, NY: Autonomedia, 2014); Nancy Fraser, *Fortunes of Feminism: From State-Managed Capitalism to Neoliberal Crisis*, Radical Thinkers (Brooklyn, NY: Verso Books, 2020); Marilyn Waring, "Counting for Something! Recognising Women's Contribution to the Global Economy through Alternative Accounting Systems," *Gender & Development* 11, no. 1 (2003): 35–43, https://doi.org/10.1080/741954251; J. K. Gibson-Graham, *The End of Capitalism (as We Knew It): A Feminist Critique of Political Economy* (Minneapolis: University of Minnesota Press, 2006).

9. Kylie Jarrett, "The Relevance of 'Women's Work': Social Reproduction and Immaterial Labor in Digital Media," *Television & New Media* 15, no. 1 (2014): 14, https://doi.org/10.1177/1527476413487607.

10. GPS refers to Global Positioning System, a system that uses triangulated satellite signals to pinpoint the location of specific devices. Most smartphones in use at the time of writing are equipped with GPS capabilities. IP geolocation refers to Internet Protocol geolocation, which is a method of approximating the geographic location of a device from its Internet Protocol address, a numerical label unique to each device connected to the internet.

11. Two notable instances of this struggle have taken place in California and in the United Kingdom. California statute AB5 was signed into law in 2019, allowing gig workers to be reclassified as employees, instead of independent contractors, expanding their protections and benefits. Soon after, however Proposition 22 passed in the November 2020 elections, after a campaign in which firms like Uber and Lyft spent hundreds of millions of dollars, Proposition 22 exempted ridesharing and delivery companies from AB5 requirements, thus allowing for the reclassification of drivers and other gig workers once again as independent contractors. At the time of writing this Proposition has been overturned by Alameda County Superior Court Judge Frank Roesch. (Suhauna Hussain, "Prop. 22 Was Ruled Unconstitutional. What Will the Final Outcome Be?," *Los Angeles Times*, August 25, 2021, https://www.latimes.com/business/technology/story/2021-08-25/after-prop-22-ruling-whats-next-uber-lyft.) Another important legal battle over labor rights in the gig economy is the UK Supreme Court ruling against Uber Technologies Inc. in February 2021, which found that "drivers should be entitled to rights such as minimum wage, holiday pay, and rest breaks." Erin Mulvaney and Kathleen Dailey, "Will Uber's U.K. Loss Jump the Pond? Gig Worker Status Explained," *Bloomberg Law*, February 22, 2021, https://news.bloomberglaw.com/daily-labor-report/will-ubers-u-k-loss-jump-the-pond-gig-worker-status-explained.

12. Such formulas can vary depending on the platform and are often described only in general terms. For instance, on their website for drivers, Grub-

hub defines its pay model through the following equation: "Mileage per order + Time spent on the road + Tips + Special offers = Total pay." However, from this material it is not clear how each variable is weighted and how much the platform itself makes from each order. Grubhub, "Getting Paid as a Grubhub Driver," Grubhub for Drivers, 2021, https://driver.grubhub.com/pay/.

13. Eric S. Sheppard, *Limits to Globalization: Disruptive Geographies of Capitalist Development* (Oxford: Oxford University Press, 2016).

14. See, for instance, Christian Berndt, Jamie Peck, and Norma M. Rantisi, eds., *Market/Place: Exploring Spaces of Exchange*, Economic Transformations (Newcastle upon Tyne: Agenda Publishing, 2020); Christian Berndt and Marc Boeckler, "Geographies of Marketization," in *The Wiley-Blackwell Companion to Economic Geography*, ed. Trevor Barnes, Jamie Peck, and Eric Sheppard (Chichester, UK: John Wiley & Sons, Ltd, 2012), 199–212, http://doi.wiley.com/10.1002/9781118384497.ch12; Christian Berndt and Marc Boeckler, "Geographies of Circulation and Exchange: Constructions of Markets," *Progress in Human Geography* 33, no. 4 (2009): 535–51, https://doi.org/10.1177/0309132509104805; Trevor J. Barnes and Brett Christophers, *Economic Geography: A Critical Introduction*, Critical Introductions to Geography (Hoboken, NJ: John Wiley & Sons, 2018); Luis F. Alvarez León, Brett Christophers, and Leqian Yu, eds., "The Spatial Constitution of Markets (Special Issue)," *Economic Geography* 94, no. 3 (2018); Brett Christophers, "The Territorial Fix: Price, Power and Profit in the Geographies of Markets," *Progress in Human Geography* 38 (January 7, 2015): 754–70, https://doi.org/10.1177/0309132513516176.

15. Some of the most influential works in this tradition are Frances C. Cairncross, *The Death of Distance: How the Communications Revolution Is Changing Our Lives*, 2nd ed. (Boston: Harvard Business Press, 1997); Thomas L. Friedman, *The World Is Flat: A Brief History of the Twenty-First Century* (New York: Farrar, Straus and Giroux, 2005). For an examination of how information technologies have transformed, rather than eliminated the role of distance (and, more generally, that of geography), see Stephen Graham, "The End of Geography or the Explosion of Place? Conceptualizing Space, Place and Information Technology," *Progress in Human Geography* 22, no. 2 (April 1, 1998): 165–85, https://doi.org/10.1191/030913298671334137; Matthew Zook and Mark Graham, "From Cyberspace to DigiPlace: Visibility in an Age of Information and Mobility," in *Societies and Cities in the Age of Instant Access*, ed. Harvey J. Miller (Dordrecht: Springer, 2007), 241–54; Agnieszka Leszczynski, "[Digital] Spatialities," in *Digital Geographies*, ed. James Ash, Rob Kitchin, and Agnieszka Leszczynski (Thousand Oaks, CA: SAGE Publications, 2018), 13–23.

16. The arguments and discussions developed in this section are adapted and expanded from chapter 1 of my doctoral dissertation Luis Felipe Alvarez Leon, "Assembling Digital Economies: Geographic Information Markets and Intellectual Property Regimes in the United States and the European Union," PhD diss., UCLA, 2016.

17. The geoweb has been defined in a variety of ways, but Leszczynski and Wilson's explanation captures the broad and transformative range of technological and social dynamics that characterize it. They refer to the geoweb as

"*both* the new materialities and practices" related to "the general convergence of location with digital information communication technologies (ICTs) [that] has brought about profound shifts in the content, forms, and practices that surround spatial media." See Agnieszka Leszczynski and Matthew W. Wilson, "Guest Editorial: Theorizing the Geoweb," *GeoJournal* 78 (July 12, 2013): 915, https://doi.org/10.1007/s10708-013-9489-7.

18. See, for instance Neil Brenner, Jamie Peck, and Nik Theodore, "After Neoliberalization?," *Globalizations* 7, no. 3 (2010): 327–45, https://doi.org/10.1080/14747731003669669; Jamie Peck, Nik Theodore, and Neil Brenner, "Neoliberal Urbanism: Models, Moments, Mutations," *SAIS Review* 29, no. 1 (2009): 49–66; A. D. Dixon, "Variegated Capitalism and the Geography of Finance: Towards a Common Agenda," *Progress in Human Geography* 35, no. 2 (2011): 193–210, https://doi.org/10.1177/0309132510372006; Bob Jessop, "Rethinking the Diversity of Capitalism: Varieties of Capitalism, Variegated Capitalism, and the World Market," in *Capitalist Diversity and Diversity within Capitalism*, ed. Geoffrey Wood and Christel Lane (London: Routledge, 2011), 209–37; Huw Macartney, *Variegated Neoliberalism: EU Varieties of Capitalism and International Political Economy* (London: Taylor & Francis, 2010); Jamie Peck and Nik Theodore, "Variegated Capitalism," *Progress in Human Geography* 31, no. 6 (2007): 731–72, https://doi.org/10.1177/0309132507083505.

19. David Harvey, *The Condition of Postmodernity: An Enquiry into the Origins of Cultural Change* (Malden, MA: Blackwell, 1989).

20. Doreen Massey, "A Global Sense of Place," in *Space, Place, and Gender* (Minneapolis: University of Minnesota Press, 1994), 149.

21. Matthew Zook, "The Geographies of the Internet," *Annual Review of Information Science and Technology* 40 (2006): 53–78, https://doi.org/10.1002/aris.1440400109; Matthew Zook et al., "New Digital Geographies: Information, Communication and Place," ed. S. D. Brunn, S. L. Cutter, and J. W. Harrington (Dordrecht, The Netherlands: Kluwer Academic Publications, 2004), 155–76.

22. Stephen Graham and Simon Marvin, *Telecommunications and the City: Electronic Spaces, Urban Places* (London: Routledge, 2002); Stephen Graham and Simon Marvin, *Splintering Urbanism: Networked Infrastructures, Technological Mobilities and the Urban Condition* (London: Routledge, 2001); Stephen Graham, "Software-Sorted Geographies," *Progress in Human Geography* 29, no. 5 (2005): 562–80, https://doi.org/10.1191/0309132505ph5680a; Stephen Graham, *Cities under Siege: The New Military Urbanism* (London: Verso, 2011).

23. Rob Kitchin and Martin Dodge, *Code/Space: Software and Everyday Life* (Cambridge, MA: MIT Press, 2011).

24. Richard Perkins and Eric Neumayer, "Is the Internet Really New After All? The Determinants of Telecommunications Diffusion in Historical Perspective," *Professional Geographer* 63 (2011): 55–72.

25. Manuel Castells, *The Internet Galaxy* (Oxford: Oxford University Press, 2001); Dieter F. Kogler, David L. Rigby, and Isaac Tucker, "Mapping Knowledge Space and Technological Relatedness in US Cities," *European Planning Studies* 21, no. 9 (2013): 1374–91, https://doi.org/10.1080/09654313.2012

.755832; Edward J. Malecki, "Digital Development in Rural Areas: Potentials and Pitfalls," *Journal of Rural Studies* 19 (2003): 201–14, https://doi.org/10 .1016/s0743-0167(02)00068-2; Edward J. Malecki and Bruno Moriset, "Organization versus Space: The Paradoxical Geographies of the Digital Economy," *Geography Compass* 3, no. 1 (2009): 256–74, https://doi.org/10.1111/j.1749 -8198.2008.00203.x; Edward J. Malecki and Bruno Moriset, *The Digital Economy: Business Organization, Production Processes and Regional Developments* (Abingdon, UK: Routledge, 2007).

26. Edward E. Leamer and Michael Storper, "The Economic Geography of the Internet Age," *NBER Working Paper Series*, no. 8450 (August 2001); Malecki and Moriset, *The Digital Economy*.

27. Peter Hall, "Creative Cities and Economic Development," *Urban Studies* 37 (1999): 639–49; Allen J. Scott, *The Cultural Economy of Cities: Essays on the Geography of Image-Producing Industries* (London: SAGE, 2000).

28. G. Grabher and O. Ibert, "Distance as Asset? Knowledge Collaboration in Hybrid Virtual Communities," *Journal of Economic Geography* 14, no. 1 (January 2014): 97–123, https://doi.org/10.1093/jeg/lbt014; Walter W. Powell and Kaisa Snellman, "The Knowledge Economy," *Annual Review of Sociology* 30 (2004): 199–220; Harald Bathelt, Anders Malmberg, and Peter Maskell, "Clusters and Knowledge: Local Buzz, Global Pipelines and the Process of Knowledge Creation," *Progress in Human Geography* 28, no. 1 (2004): 31–56, https://doi.org/10.1191/0309132504ph469oa; M. Lorenzen and R. Mudambi, "Clusters, Connectivity and Catch-up: Bollywood and Bangalore in the Global Economy," *Journal of Economic Geography* 13, no. 3 (May 2013): 501–34, https://doi.org/10.1093/jeg/lbs017; M. Porter, "Location, Competition, and Economic Development: Local Clusters in a Global Economy," *Economic Development Quarterly* 14 (2000): 15–34.

29. C. Gibson and L. Kong, "Cultural Economy: A Critical Review," *Progress in Human Geography* 29, no. 5 (2005): 541–61; Dominic Power and Allen J. Scott, *Cultural Industries and the Production of Culture* (New York: Psychology Press, 2004); Allen J. Scott, *On Hollywood: The Place, The Industry* (Princeton: Princeton University Press, 2005).

30. A. J. Scott, "Capitalism and Urbanization in a New Key? The Cognitive-Cultural Dimension," *Social Forces* 85, no. 4 (2007): 1465–82; Allen J. Scott, *Social Economy of the Metropolis: Cognitive-Cultural Capitalism and the Global Resurgence of Cities* (Oxford University Press, 2008); Paul Langley and Andrew Leyshon, "Platform Capitalism: The Intermediation and Capitalization of Digital Economic Circulation," *Finance and Society* 3, no. 1 (2017): 11–31, https://doi.org/10.2218/finsoc.v3i1.1936; Nick Srnicek, *Platform Capitalism* (Cambridge, UK: Polity Press, 2017); Malecki and Moriset, *The Digital Economy*; Shoshana Zuboff, *The Age of Surveillance Capitalism: The Fight for a Human Future at the New Frontier of Power*, First Trade Paperback Edition (New York: PublicAffairs, 2020); Elizabeth Currid, *The Warhol Economy* (Princeton, NJ: Princeton University Press, 2007).

31. Deborah Leslie and Norma M. Rantisi, "Creativity and Place in the Evolution of a Cultural Industry the Case of Cirque Du Soleil," *Urban Studies* 48, no. 9 (July 2011): 1771–87; Hiro Izushi and Yuko Aoyama, "Industry Evolution

and Cross-Sectoral Skill Transfers: A Comparative Analysis of the Video Game Industry in Japan, the United States, and the United Kingdom," *Environment and Planning A: Economy and Space* 38 (2006): 1843–61, https://doi.org/10.1068/a37205; Brett Christophers, *Envisioning Media Power: On Capital and Geographies of Television* (Lanham, MD: Lexington Books, 2009); Luis F. Alvarez León, "The Emergence of Netflix and the New Digital Economic Geography of Hollywood," in *Netflix at the Nexus: Content, Practice, and Production in the Age of Streaming Television*, ed. Amber M. Buck and Theo Plothe (New York: Peter Lang, 2019), 47–63; Elizabeth Currid-Halkett and Allen J. Scott, "The Geography of Celebrity and Glamour: Reflections on Economy, Culture, and Desire in the City," *City, Culture and Society*, 2013; H. Yoon and E. J. Malecki, "Cartoon Planet: Worlds of Production and Global Production Networks in the Animation Industry," *Industrial and Corporate Change* 19, no. 1 (February 2010): 239–71, https://doi.org/10.1093/icc/dtp040; Glen Norcliffe and Olivero Rendace, "New Geographies of Comic Book Production in North America: The New Artisan, Distancing, and the Periodic Social Economy," *Economic Geography* 79, no. 3 (2003): 241–63, https://doi.org/10.1111/j.1944-8287.2003.tb00211.x; Currid, *Warhol Economy*.

32. Rob Kitchin, *The Data Revolution* (Los Angeles: SAGE, 2014).

33. Arno Scharl and Klaus Tochtermann, *The Geospatial Web* (London: Springer, 2007).

34. Tim O'Reilly, "What Is Web 2.0: Design Patterns and Business Models for the Next Generation of Software," *Communications and Strategies* 65, no. 1 (January 1, 2007): 17–37.

35. Michael F. Goodchild, "Citizens as Sensors: Web 2.0 and the Volunteering of Geographic Information," *GeoFocus*, no. 69 (2007): 211–21.

36. See for example the edited volume Daniel Sui, Sarah Elwood, and Michael Goodchild, *Crowdsourcing Geographic Knowledge* (Dordrecht: Springer, 2013). Among the many applications and issues covered by the contributions in this volume, a representative sample includes: Marcus Goetz and Alexander Zipf, "The Evolution of Geo-Crowdsourcing: Bringing Volunteered Geographic Information to the Third Dimension," in *Crowdsourcing Geographic Knowledge*, ed. Daniel Sui, Sarah Elwood, and Michael F. Goodchild (Dordrecht: Springer, 2013), 139–58; Muki Haklay, "Citizen Science and Volunteered Geographic Information: Overview and Typology of Participation," ed. Daniel Sui, Sarah Elwood, and Michael F. Goodchild (Dordrecht: Springer, 2013), 105–20; Peter A. Johnson and Renee E. Sieber, "Situating the Adoption of VGI by Government," ed. Daniel Sui, Sarah Elwood, and Michael F. Goodchild (Dordrecht: Springer, 2013), 65–81; Mark H. Palmer and Scott K. Kraushaar, "Volunteered Geographic Information, Actor-Network Theory, and Severe-Storm Reports," ed. Daniel Sui, Sarah Elwood, and Michael F. Goodchild (Dordrecht: Springer, 2013), 287–306; Christopher Goranson, Sayone Thihalolipavan, and Nicolás di Tada, "VGI and Public Health: Possibilities and Pitfalls" (Dordrecht: Springer Science & Business Media, 2013), 329–40.

37. Barney Warf and Daniel Sui, "From GIS to Neogeography: Ontological Implications and Theories of Truth," *Annals of GIS* 16, no. 4 (2010): 197–209, https://doi.org/10.1080/19475683.2010.539985.

38. Andrew Turner, *Introduction to Neogeography* (Sebastopol, CA: O'Reilly Media, Inc., 2006).

39. Matthew W. Wilson, "Situating Neogeography," *Environment and Planning A* 45, no. 1 (2013): 3–9; Matthew W. Wilson and Mark Graham, "Neogeography and Volunteered Geographic Information: A Conversation with Michael Goodchild and Andrew Turner," *Environment and Planning A* 45, no. 1 (2013): 10–18, https://doi.org/10.1068/a44483; Jeremy W. Crampton et al., "Beyond the Geotag: Situating 'Big Data' and Leveraging the Potential of the Geoweb," *Cartography and Geographic Information Science* 40, no. 2 (2013): 130–39, https://doi.org/10.1080/15230406.2013.777137.

40. Tanmoy Das, C. P. J. M. van Elzakker, and Menno-Jan Kraak, "Conflicts in Neogeography Maps," in *Proceedings—AutoCarto 2012*, Columbus, OH, September 16–18, 2012.

41. Mordechai Muki Haklay, "Neogeography and the Delusion of Democratisation," *Environment and Planning A* 45, no. 1 (2013): 55–69, https://doi.org/10.1068/a45184.

42. Yochai Benkler, Rob Faris, and Hal Roberts, *Network Propaganda: Manipulation, Disinformation, and Radicalization in American Politics* (New York: Oxford University Press, 2018); Zeynep Tufekci, *Twitter and Tear Gas: The Power and Fragility of Networked Protest* (New Haven, CT: Yale University Press, 2017); Tim Wu, *The Master Switch: The Rise and Fall of Information Empires* (New York: Vintage Books, 2011); Tim Wu, *The Attention Merchants: The Epic Scramble to Get Inside Our Heads* (New York: Vintage Books, a division of Penguin Random House LLC, 2017).

43. Sarah Elwood and Agnieszka Leszczynski, "Privacy, Reconsidered: New Representations, Data Practices, and the Geoweb," *Geoforum* 42, no. 1 (January 2011): 6–15, https://doi.org/10.1016/j.geoforum.2010.08.003; Mark Graham, "Time Machines and Virtual Portals: The Spatialities of the Digital Divide," *Progress in Development Studies* 11, no. 3 (2011): 211–27; Jeremy Crampton, "Collect It All: National Security, Big Data and Governance," *GeoJournal* 80 (2015): 519–31, https://doi.org/10.1007/s10708-014-9598-y; Michael Goodchild et al., "A White Paper on Locational Information and the Public Interest" (American Association of Geographers, September 2022), https://doi.org/10.14433/2017.0113.

44. Kitchin, *The Data Revolution*.

45. Catherine D'Ignazio and Lauren F. Klein, *Data Feminism*, Strong Ideas Series (Cambridge, MA: The MIT Press, 2020).

46. Craig M. Dalton, Linnet Taylor, and Jim Thatcher, "Critical Data Studies: A Dialog on Data and Space," *Big Data & Society* 3, no. 1 (2016), https://doi.org/10.1177/2053951716648346; Ryan Burns, Craig M. Dalton, and Jim E. Thatcher, "Critical Data, Critical Technology in Theory and Practice," *The Professional Geographer* 70, no. 1 (2018): 126–28, https://doi.org/10.1080/00330124.2017.1325749; Craig Dalton and Jim Thatcher, "What Does a Critical Data Studies Look Like, and Why Do We Care? Seven Points for a Critical Approach to 'Big Data,'" Society and Space open site, May 12, 2014, https://www.societyandspace.org/articles/what-does-a-critical-data-studies-look-like-and-why-do-we-care; Rob Kitchin and Tracey P. Lauriault, "Towards Critical

Data Studies: Charting and Unpacking Data Assemblages and Their Work," *The Programmable City Working Paper*, 2014, http://papers.ssrn.com/sol3/papers .cfm?abstract_id=2474112.

47. See, for instance Kitchin and Lauriault, "Towards Critical Data Studies"; Dalton, Taylor, and Thatcher, "Critical Data Studies; danah boyd and Kate Crawford, "Critical Questions for Big Data," *Information, Communication and Society* 15, no. 5 (2012): 662–79, https://doi.org/10.1080/1369118X.2012.678878; Burns et al., "Critical Data, Critical Technology"; Dalton and Thatcher, "What Does a Critical Data Studies Look Like?"

48. John Pickles, *Ground Truth: The Social Implications of Geographic Information Systems* (New York: Guilford Press, 1995).

49. boyd and Crawford, "Critical Questions for Big Data"; Sarah Elwood, "Volunteered Geographic Information: Future Research Directions Motivated by Critical, Participatory, and Feminist GIS," *GeoJournal* 72 (2008): 173–83, https://doi.org/10.1007/s10708-008-9186-0; Crampton et al., "Beyond the Geotag"; M. Graham and T. Shelton, "Geography and the Future of Big Data, Big Data and the Future of Geography," *Dialogues in Human Geography* 3, no. 3 (2013): 255–61, https://doi.org/10.1177/2043820613513121; Leszczynski and Wilson, "Guest Editorial"; M-P. Kwan, "Feminist Visualization: Re-Envisioning GIS as a Method in Feminist Geographic Research," *Annals of the Association of American Geographers* 92, no. 4 (2002): 645–61; Nadine Schuurman and Geraldine Pratt, "Care of the Subject: Feminism and Critiques of GIS," *Gender, Place & Culture* 9, no. 3 (2002): 291–99, https://doi.org/10 .1080/0966369022000003905; Nadine Schuurman, "Trouble in the Heartland: GIS and Its Critics in the 1990s," *Progress in Human Geography* 24, no. 4 (2000): 569–90; Pickles, *Ground Truth*; Eric Sheppard, "Knowledge Production through Critical GIS: Genealogy and Prospects," *Cartographica: The International Journal for Geographic Information* 40, no. 4 (Winter 2005): 5–21; Dalton et al., "Critical Data Studies"; Burns, et al., "Critical Data, Critical Technology"; D'Ignazio and Klein, *Data Feminism*.

50. Harrison Smith, "Open and Free? The Political Economy of the Geospatial Web 2.0," *Geothink Working Paper Series*, Geothink Working Paper Series, no. 001 (2014), http://geothink.ca/wp-content/uploads/2014/06/Geothink -Working-Paper-001-Shade-Smith1.pdf; Agnieszka Leszczynski, "Situating the Geoweb in Political Economy," *Progress in Human Geography* 36, no. 1 (2012): 72–89, https://doi.org/10.1177/0309132511411231; Luis F. Alvarez León, "Information Policy and the Spatial Constitution of Digital Geographic Information Markets," *Economic Geography* 94, no. 3 (2018): 217–37, https://doi .org/10.1080/00130095.2017.1388161.3.

CHAPTER 2. ASSEMBLING THE BASE MAP

1. With the notable, and growing exception of web maps, which are designed specifically to be accessed, and used, online.

2. As will be explained in greater detail in chapter 3, the "location of the query" masks a much more complex set of spatial relations, which can include,

for instance, the location about which the query is made, the location of the user at the moment of making the query, or the location associated with the device through which the query is made, among others. The location of a query is in turn also shaped by infrastructural, economic, and political factors, from cell phone carrier coverage and the spatial distribution of communications infrastructure (e.g., cellular towers), to content distribution agreement and even government censorship, all of which interact to shape who can access what, where, and how.

3. While ground beef has been present in many cultures and epochs, there is some consensus that the modern burger can be dated to nineteenth-century German cuisine, especially the "Hamburg steak," and was brought to the United States by waves of German immigrants. It was in this new cultural milieu that the hamburger was modified, adapting to local tastes and influences. Through this transformation emerged the dish that would become both a symbol of United States culture (and cultural imperialism), as well as the paradigmatic product of the assembly-line preparation that undergirded the fast-food industry as pioneered by McDonald's and spread globally by other multinational corporations like Burger King and White Castle. For a more thorough historical treatment, see Josh Ozersky, *The Hamburger: A History* (New Haven, CT: Yale University Press, 2008).

4. Kurt Iveson and Sophia Maalsen, "Social Control in the Networked City: Datafied Dividuals, Disciplined Individuals and Powers of Assembly," *Environment and Planning D: Society and Space* 37, no. 2 (2019): 331–49, https://doi.org/10.1177/0263775818812084.

5. Agnieszka Leszczynski, "Spatial Big Data and Anxieties of Control," *Environment and Planning D: Society and Space* 33, no. 6 (2015): 965–84, https://doi.org/10.1177/0263775815595814.

6. Jovanna Rosen and Luis F. Alvarez León, "Signaling Hinterlands and the Spatial Networks of Digital Capitalism," *Annals of the American Association of Geographers* (2023): 1-13, https://doi.org/10.1080/24694452.2023.2249974.

7. See Patricia Callejo et al., "A Deep Dive into the Accuracy of IP Geolocation Databases and Its Impact on Online Advertising," *IEEE Transactions on Mobile Computing* 22, no. 8 (2022): 1–15, https://doi.org/10.1109/TMC.2022.3166785.

8. University of Chicago Library, "Guide to the R.R. Donnelley & Sons Company Archive 1844–2005," https://www.lib.uchicago.edu/e/scrc/findingaids/view.php?eadid=ICU.SPCL.DONNELLEY.

9. Barbara B. Petchenik, "Donnelley Cartographic Services," *American Cartographer* 14, no. 3 (January 1987): 241–44, https://doi.org/10.1559/15230408 7783875796.

10. Joe Francica, "MapQuest Technology and Applications," *Directions Magazine*, March 26, 2004, 6.

11. Francica, 6.

12. Francica, 6.

13. Chico Harlan, "'Does MapQuest Still Exist?' Yes, It Does, and It's a Profitable Business," *Washington Post*, May 22, 2015, sec. Business, https://www

.washingtonpost.com/business/economy/does-mapquest-still-exist-as-a-matter-of
-fact-it-does/2015/05/22/995d2532-fa5d-11e4-a13c-193b1241d51a_story.html.

14 J. B. Harley, "Deconstructing the Map," *Cartographica: The International Journal for Geographic Information and Geovisualization* 26 no. 2 (Summer 1989): 1–20, https://doi.org/10.3138/E635-7827-1757-9T53. For a broader survey of these arguments, and their role in the more expansive conversations undergirding the development of Critical GIS and Critical Cartography, see Jeremy W. Crampton, *Mapping: A Critical Introduction to Cartography and GIS*, Critical Introductions to Geography (Malden, MA: Wiley-Blackwell, 2010); Jeremy W. Crampton, "Mappings," *The Wiley-Blackwell Companion to Cultural Geography* (Malden, MA: Wiley-Blackwell, 2013), 423–36; Matthew W. Wilson, "New Lines? Enacting a Social History of GIS," *Canadian Geographer / Le Géographe Canadien* 59, no. 1 (2015): 29–34, https://doi.org/10.1111/cag.12118.

15. I take up this discussion at length through the analysis of the various property regimes governing Google Maps in my article Luis F. Alvarez León, "Property Regimes and the Commodification of Geographic Information: An Examination of Google Street View," *Big Data & Society* 3, no. 2 (2016): 2053951716637885, https://doi.org/10.1177/2053951716637885.

16. Michael P. Peterson, "MapQuest and the Beginnings of Web Cartography," *International Journal of Cartography* 7, no. 2 (2021): 275–81, https://doi.org/10.1080/23729333.2021.1925831.

17. Matthew Crain, *Profit over Privacy: How Surveillance Advertising Conquered the Internet* (Minneapolis: University of Minnesota Press, 2021), 1.

18. Peterson, "MapQuest and the Beginnings of Web Cartography."

19. Peterson, "MapQuest and the Beginnings of Web Cartography," 280.

20. Soon after, in 2008, Google launched the Google Maps application for Android, which would coincide with the release of the first commercial device to use Google's Android operating system (the HTC Dream smartphone). In its 2021 Google I/O Keynote the company announced that as of May 2021 there were over three billion active Android devices in the world—the most of any mobile operating system. Furthermore, one billion of those devices were added in the previous year. See Ben Schoon, "Numbers from Google I/O: 3x Growth for Wear OS, 3 Billion Active Android Devices, More," *9to5Google* (blog), May 12, 2022, https://9to5google.com/2022/05/11/google-io-2022-numbers/.

21. Sergey Brin and Lawrence Page, "The Anatomy of a Large-Scale Hypertextual Web Search Engine," *Computer Networks and ISDN Systems*, Proceedings of the Seventh International World Wide Web Conference, 30, no. 1 (April 1998): 107–17, https://doi.org/10.1016/S0169-7552(98)00110-X.

22. See Google, "Our Approach—How Google Search Works," accessed June 17, 2022, https://www.google.com/search/howsearchworks/our-approach/.

23. Adam Fisher, "Google's Road Map to Global Domination," *New York Times Magazine*, December 11, 2013, http://www.nytimes.com/2013/12/15/magazine/googles-plan-for-global-domination-dont-ask-why-ask-where.html?_r=0.

24. Google Earth was a particularly controversial case due to claims that Keyhole, the company that developed what would become Google Earth, had

stolen code from German collective ART+COM, and their product Terravision, which also provided a virtual representation of Planet Earth. This would lead to a $100 million patent infringement lawsuit filed by ART+COM against Google. Eventually ART+COM lost the lawsuit, and its patent was invalidated. This saga was the basis for the Netflix series *The Billion Dollar Code*. For more on the development of Google Earth and its controversies, see Jeremy W. Crampton, "Keyhole, Google Earth, and 3D Worlds: An Interview with Avi Bar-Zeev," *Cartographica: The International Journal for Geographic Information and Geovisualization* 43, no. 2 (Summer 2008): 85–93, https://doi.org/10.3138/carto.43.2.85; Avi Bar-Zeev, "Was Google Earth Stolen? (No)," *Medium* (blog), October 7, 2022, https://avibarzeev.medium.com/was-google-earth-stolen-7d1b821e589b.

25. For a full list and documentation of all of Google's acquisitions (259 at the time of writing), see Crunchbase, "List of Google's 259 Acquisitions, Including Photomath and Alter," Crunchbase, accessed July 30, 2023, https://www.crunchbase.com/search/acquisitions/field/organizations/num_acquisitions/google.

26. For discussions on how street-level imagery is collected, its various applications, the different forms of value it can take, and its distribution throughout cities all over the world, see my articles with Sterling Quinn on this topic: Sterling Quinn and Luis F. Alvarez León, "Every Single Street? Rethinking Full Coverage across Street-Level Imagery Platforms," *Transactions in GIS*, 23, no. 6 (2019): 1251–72, https://doi.org/10.1111/tgis.12571; Luis Alvarez Leon and Sterling Quinn, "The Value of Crowdsourced Street-Level Imagery: Examining the Shifting Property Regimes of OpenStreetCam and Mapillary," *GeoJournal* 84, no. 2 (April 2019): 395–414, https://doi.org/10.1007/s10708-018-9865-4; Luis Alvarez Leon and Sterling Quinn, "Street-Level Imagery," *Geographic Information Science & Technology Body of Knowledge* 2019, no. Q4 (2019), https://doi.org/10.22224/gistbok/2019.4.12.

CHAPTER 3. LOCATION, GEOLOCATION, ALLOCATION

1. Following political scientist Harold Lasswell's influential definition of politics, developed in his 1936 book with the same subtitle: Harold Dwight Lasswell, *Politics: Who Gets What, When, How* (Whitefish, MT: Literary Licensing, 2013).

2. Stephen Tung, "How the Finnish School System Outshines US Education," Stanford Report, January 20, 2012, http://news.stanford.edu/news/2012/january/finnish-schools-reform-012012.html.

3. See "Queering The Map," accessed July 14, 2022, https://www.queeringthemap.com/. For a rich investigation focusing on the specific locations that underpinned and nurtured development of the queer community in New York City, see the work of Jen Jack Gieseking, specifically the book *A Queer New York: Geographies of Lesbians, Dykes, and Queers* (New York: New York University Press, 2020). and the related digital interactive map "An Everyday Queer New York," http://jgieseking.org/AQNY/AEQNYpubsmap/index.html.

4. Trevor J. Barnes, "'Desk Killers': Walter Christaller, Central Place Theory, and the Nazis," in *Geographies of Knowledge and Power*, ed. Peter Meusburger,

Derek Gregory, and Laura Suarsana (Dordrecht: Springer, 2015), 187, https://doi.org/10.1007/978-94-017-9960-7_9. For more on the history of Central Place Theory and other locational models and their key role in Nazi "reterritorialization," see Trevor J. Barnes and Claudio Minca, "Nazi Spatial Theory: The Dark Geographies of Carl Schmitt and Walter Christaller," *Annals of the Association of American Geographers* 103 (2013): 669–87, https://doi.org/10.2307/23485411?ref=no-x-route:2ca919d517fd198be3b5a9b6c83ce18a; Trevor J. Barnes, "Reopke Lecture in Economic Geography: Notes from the Underground: Why the History of Economic Geography Matters: The Case of Central Place Theory," *Economic Geography* 88, no. 1 (2012): 1–26, https://doi.org/10.1111/j.1944-8287.2011.01140.x.

5. Trevor Barnes, "A Morality Tale of Two Location Theorists in Hitler's Germany: Walter Christaller and August Lösch," in *Hitler's Geographies: The Spatialities of the Third Reich*, ed. Paolo Giaccaria and Claudio Minca (Chicago: University of Chicago Press, 2016), 212.

6. For a historical perspective of this and other transitions in economic geography in the second half of the twentieth century, see A. J. Scott, "Economic Geography: The Great Half-Century," *Cambridge Journal of Economics* 24, no. 4 (2000): 483–504.

7. While this turn involved many participants and cannot be reduced to a single actor or explanation, as a shorthand it is often exemplified through the sharp pivot that marked the academic career of prominent economic geographer David Harvey. He began his career as a quantitatively oriented scholar, publishing a landmark volume, *Explanation in Geography*, in 1969. However, after moving from Great Britain to the United States, his experiences witnessing racial and social inequality and urban upheaval in Baltimore in the late 1960s led him away from quantitative geography (which he saw as largely unconcerned with such social phenomena) and toward a more qualitative, decidedly Marxist approach, publishing the volume *Social Justice and the City* in 1973, and many later volumes focusing on analyzing and critiquing the spatial dimensions of capitalism, work which fueled a broader disciplinary shift in economic geography. For an autobiographical reflection documenting this "epistemic break" in Harvey's trajectory and its relationship to broader historical developments in geography, see David Harvey, "Reflections on an Academic Life," *Human Geography* 15, no. 1 (2022): 14–24, https://doi.org/10.1177/19427786211046291.

8. For a glimpse into these debates see, among others, Stan Openshaw, "A View on the GIS Crisis in Geography, or, Using GIS to Put Humpty-Dumpty Back Together Again," *Environment and Planning A* 23, no. 5 (1991): 621–28; Stan Openshaw, "The Truth about Ground Truth," *Transactions in GIS* 2, no. 1 (1997): 7–24; Nadine Schuurman, "Trouble in the Heartland: GIS and Its Critics in the 1990s," *Progress in Human Geography* 24, no. 4 (2000): 569–90; Neil Smith, "History and Philosophy of Geography: Real Wars, Theory Wars," *Progress in Human Geography* 16, no. 2 (1992): 257–71; Sarah Elwood, Nadine Schuurman, and Matthew W. Wilson, "Critical GIS," in *The SAGE Handbook of GIS and Society*, ed. Timothy L. Nyerges, Helen Couclelis, and Robert McMaster (London: SAGE, 2011), 87–106, https://doi.org/10.4135/9781446201046.n5; M-P. Kwan, "Feminist Visualization: Re-Envisioning GIS

as a Method in Feminist Geographic Research," *Annals of the Association of American Geographers* 92, no. 4 (2002): 645–61; Mei-Po Kwan, "Introduction: Feminist Geography and GIS," *Gender, Place & Culture* 9, no. 3 (2002): 261–62, https://doi.org/10.1080/0966369022000003860; John Pickles, *Ground Truth: The Social Implications of Geographic Information Systems* (New York: Guilford Press, 1995); John Pickles, "Arguments, Debates, and Dialogues: The GIS-Social Theory Debate and the Concern for Alternatives," *Geographical Information Systems* 1 (1999): 49–60; Eric Sheppard, "Knowledge Production through Critical GIS: Genealogy and Prospects," *Cartographica: The International Journal for Geographic Information* 40, no. 4 (Winter 2005): 5–21.

9. For some examples of this body of work, Pierre-Alexandre Balland and David Rigby, "The Geography of Complex Knowledge," *Economic Geography* 93, no. 1 (2017): 1–23, https://doi.org/10.1080/00130095.2016.1205947; Sergio Petralia, Pierre-Alexandre Balland, and David L. Rigby, "Unveiling the Geography of Historical Patents in the United States from 1836 to 1975," *Scientific Data* 3 (2016): 160074, https://doi.org/10.1038/sdata.2016.74; Ron Boschma, Pierre-Alexandre Balland, and Dieter Franz Kogler, "Relatedness and Technological Change in Cities: The Rise and Fall of Technological Knowledge in US Metropolitan Areas from 1981 to 2010," *Industrial and Corporate Change* 24, no. 1 (February 2015): 223–50, https://doi.org/10.1093/icc/dtu012; Pierre-Alexandre Balland et al., "Smart Specialization Policy in the European Union: Relatedness, Knowledge Complexity and Regional Diversification," *Regional Studies* 53, no. 9 (2019): 1252–68, https://doi.org/10.1080/00343404.2018.1437900.

10. This question is at the heart of Mike Davis's analysis and critique of Los Angeles as a city, as a form of urban life, and as spatial expression of twentieth-century capitalism. See in particular his two volumes dedicated to examining this city from multiple perspectives: Mike Davis, *City of Quartz: Excavating the Future in Los Angeles* (London: Verso, 2018); Mike Davis, *Ecology of Fear: Los Angeles and the Imagination of Disaster* (London: Verso, 2022).

11. The paradigmatic work documenting Chicago's rise as a gateway to the "Great West" is William Cronon, *Nature's Metropolis: Chicago and the Great West*, 3rd ed. (New York: W. W. Norton & Company, 1992).

12. The border was open to European immigrants. At different times throughout its history, the United States has had laws and rules that restrict immigration depending on factors such as race (the 1790 Naturalization Act excluded non-white people from eligibility to naturalize), and national origin (such as the Chinese Exclusion Act of 1882). To explore how these legal restrictions relate to immigration to the United States, see D'vera Cohn, "How US Immigration Laws and Rules Have Changed through History," Pew Research Center, September 2015, https://www.pewresearch.org/short-reads/2015/09/30/how-u-s-immigration-laws-and-rules-have-changed-through-history/.

13. Ran Abramitzky, Leah Platt Boustan, and Katherine Eriksson, "A Nation of Immigrants: Assimilation and Economic Outcomes in the Age of Mass Migration," *Journal of Political Economy* 122, no. 3 (2014): 467–468, https://doi.org/10.1086/675805.

14. "Irish Emigration History," University College Cork, accessed May 22, 2023, https://www.ucc.ie/en/emigre/history/#_ftn8.

15. Dylan Shane Connor, "The Cream of the Crop? Geography, Networks, and Irish Migrant Selection in the Age of Mass Migration," *Journal of Economic History* 79, no. 1 (2019): 139, https://doi.org/10.1017/S0022050718000682.

16. The World Bank, "Net Migration—Ireland," World Bank Open Data, 2022, https://data.worldbank.org.

17. The dominance of Los Angeles in global Film and TV markets is enduring, but not uncontested. While it has developed a rich repository of local assets over the past century, this location also faces competition from other cities and regions around the world, as well as changing consumption habits under globalization, and technological innovations. For an in-depth treatment of the economic geography of Hollywood, and its relationship to its location in Los Angeles, see Scott, *On Hollywood: The Place, the Industry*; Michael Storper and Susan Christopherson, "Flexible Specialization and Regional Industrial Agglomerations: The Case of the US Motion Picture Industry," *Annals of the Association of American Geographers* 77, no. 1 (1987): 104–17; Susan Christopherson and Michael Storper, "The City as Studio, the World as Backlot: The Impact of Vertical Disintegration on the Location of the Motion-Picture Industry," *Environment and Planning D: Society and Space* 4, no. 3 (1986): 305–20. For a discussion of how the emergence of streaming services and other digital technologies places Los Angeles and Hollywood in simultaneous collaboration and (increasingly) competition with Silicon Valley and the technology industry, see my previously published chapter: Alvarez León, "The Emergence of Netflix and the New Digital Economic Geography of Hollywood," in *Netflix at the Nexus: Content, Practice, and Production in the Age of Streaming Television*, ed. Amber M. Buck and Theo Plothe (New York: Peter Lang, 2019), 47–63.

18. While it is impossible to recount the astonishing variation of maps across societies, times, and places, the History of Cartography project has come the closest of any endeavor. This project is hosted by the University of Wisconsin–Madison's Department of Geography and has at its core a six-volume book series. The project was started by cartographers J. B. Harley and David Woodward in 1977, with the first volume published in 1987. The series is edited by a team of scholars, and has been published in paper and online (the first three volumes are freely accessible as PDFs) by the University of Chicago Press. For more information, see University of Wisconsin Department of Geography, "History of Cartography Project," accessed July 22, 2022, https://geography.wisc.edu/histcart/.

19. See Agnieszka Leszczynski, "Situating the Geoweb in Political Economy," *Progress in Human Geography* 36, no. 1 (2012): 72, https://doi.org/10.1177/0309132511411231.

20. For more on the history of GIS, its multiple developmental paths, and the reconceptualization of geography that they catalyzed, see Matthew W. Wilson, *New Lines: Critical GIS and the Trouble of the Map* (Minneapolis: University of Minnesota Press, 2017); Matthew W. Wilson, "New Lines? Enacting a Social History of GIS," *The Canadian Geographer / Le Géographe Canadien* 59, no. 1 (2015): 29–34, https://doi.org/10.1111/cag.12118.

21. Michael F. Goodchild, "Geographical Information Science," *International Journal of Geographical Information Systems* 6, no. 1 (1992): 31–45, https://doi.org/10.1080/02693799208901893.

22. I want to make space here for the possibility of multiple ways of thinking and doing GIS that do not necessarily fall within these parameters. One instance of this is mapping phenomena whose concrete location is not exactly known, or whose nature make it difficult or even impossible to pinpoint to specific locations, like emotions, myths, or narratives. Another is the fact that GIS can be used to map and explore spaces that are not "in the real world"—from fictional worlds to theoretical explorations—and therefore cannot satisfy the definition above. For some examples of creative, imaginative, and non-standard approaches to GIS and mapping, see the Special Issue of *The Cartographic Journal* "Cartographies of Fictional Worlds" 48, no. 4 (2011) , which includes articles such as Eva Erdmann, "Topographical Fiction: A World Map of International Crime Fiction," 274–84, https://doi.org/10.1179/1743277411Y .0000000027; Ina Habermann and Nikolaus Kuhn, "'Sustainable Fictions': Geographical, Literary and Cultural Intersections in J. R. R. Tolkien's *The Lord of the Rings*," *Cartographic Journal* 48, no. 4 (2011): 263–73, https://doi.org /10.1179/1743277411Y.0000000024; Annika Richterich, "Cartographies of Digital Fiction: Amateurs Mapping a New Literary Realism," 237–49, https:// doi.org/10.1179/1743277411Y.0000000021; and Sébastien Caquard, "Cartographies of Fictional Worlds: Conclusive Remarks," 224–25, https://doi.org /10.1179/000870411X13203362557264. For tools designed for the express purpose of reimagining cartographic spaces and rethinking how we can represent and analyze them, see Luke Bergmann and Nick Lally, "For Geographical Imagination Systems," *Annals of the American Association of Geographers* 111, no. 1 (2021): 26–35, https://doi.org/10.1080/24694452.2020.1750941.

23. In most applications, the frame of reference refers to Planet Earth, although this need not be the case. Applications such as space exploration and astronomical observation must work with other planetary bodies, while Computer Tomography uses the human body as a frame of reference. These applications use spatial data, but since they use a different frame of reference they are often discussed as separate from those *geographic* information applications that use the Earth as a frame of reference. Yet, substantively, all these cases can be thought of as subsets of spatial information.

24. US Geological Survey, "What Does 'Georeferenced' Mean?," accessed July 25, 2022, https://www.usgs.gov/faqs/what-does-georeferenced-mean. For more on georeferencing, and the related process or georectification, which entails "the removal of geometric distortions between sets of data points" see Christopher Lippitt, "Georeferencing and Georectification," *Geographic Information Science & Technology Body of Knowledge*, no. Q3 (2020), https://doi .org/10.22224/gistbok/2020.3.3.

25. Some media theorists have argued that location is in fact instrumental to the construction of digital technologies in the first place. As explored by Ranjodh Singh Dhaliwal, "addressability" is a core feature of not only the digital, but of other forms of computation as well: "addressability—the condition whereby addressing, or the practice of giving a locational/spatial index takes place—undergirds all computing as we know [it]." Ranjodh Singh Dhaliwal, "On Addressability, or What Even Is Computation?," *Critical Inquiry* (2022): 5, https://doi.org/10.1086/721167. The locational elements and spatial dimensions

of the digital in turn have broader political implications. On this point, Bernard Dionysius Geoghegan argues that the development of computerized images realizes a "reciprocal relation between writing and space, in which emerging technical standards mediate concrete geopolitical conditions, invested with specific spatial-historical conditions. The manner in which these scenes [WWII-era radar images, the development of JPEG, and the development of EXIF metadata] specified problems like resolution and interoperability reflect problems of addressability that condition the formats of electronic images. These operations locate images in an irreducibly historical and geographic orientation, which the image encodes, modulates, and reproduces. As such, these and other media formats do not merely specify the appearance of this or that image. Rather, they take part in a political ordering of place users, images, instruments, and space. The bitmap is the territory." Bernard Dionysius Geoghegan, "The Bitmap Is the Territory: How Digital Formats Render Global Positions," *MLN* 136, no. 5 (2021): 1098, https://doi.org/10.1353/mln.2021.0081.

26. Paul Baran, "On Distributed Communications: I. Introduction to Distributed Communications Networks" (RAND Corporation, August 1964), v, https://www.rand.org/pubs/research_memoranda/RM3420.html.

27. For a more comprehensive treatment of the evolution of the internet, and the various computer networks that preceded it, see Janet Abbate, *Inventing the Internet* (Cambridge, MA: MIT Press, 2000).

28. The United States was far from the only country that developed networked computing initiatives in the second half of the twentieth century. In fact, countries like the Soviet Union, France, and Chile made headway in establishing computer networks throughout their territories. Each of these cases followed different political imperatives and institutional configurations, with varying degrees of success. While ultimately it was the US network that became the basis for the internet, recent scholarship has reappraised many of the technical innovations, architectural features, and possibilities contained within each of these alternative networks. This scholarship has taken on contemporary salience in part due to the urgent problems that plague the contemporary internet, from the outsize influence of corporations to misinformation and its implications for democracy. For some key works in this area, see Eden Medina, "Designing Freedom, Regulating a Nation: Socialist Cybernetics in Allende's Chile," *Journal of Latin American Studies* 38, no. 3 (2006): 571–606, doi:10.1017/S0022216X06001179; Eden Medina, *Cybernetic Revolutionaries: Technology and Politics in Allende's Chile* (Cambridge, MA: MIT Press), 2011; Benjamin Peters, *How Not to Network a Nation: The Uneasy History of the Soviet Internet* (Cambridge, MA: MIT Press, 2017); Julien Mailland and Kevin Driscoll, *Minitel: Welcome to the Internet*, Platform Studies (Cambridge, MA: MIT Press, 2017).

29. William Gibson, *Neuromancer* (Penguin, 2000), 51.

30. Fred Turner, *From Counterculture to Cyberculture: Stewart Brand, the Whole Earth Network, and the Rise of Digital Utopianism* (Chicago: University of Chicago Press, 2008).

31. All excerpts in this paragraph are quoted from John Perry Barlow, "A Declaration of the Independence of Cyberspace," Electronic Frontier Foundation, 1996, https://www.eff.org/cyberspace-independence.

32. While the idea of the digital divide led to much research, media, and policy work in the 2000s focused on addressing the connectivity gaps and bringing the benefits of digital technologies to underserved populations, it also conflated connectedness with development, social mobility, and economic opportunity. This conflation has drawn much scrutiny from critics who argue that the focus on technology underplays the role of more fundamental social structures and policies while serving a market-centric ideology and reducing complex social problems to simple technological solutions. For a critical examination of this "access doctrine" as it transformed public goods like schools and libraries in the United States, see Daniel Greene, *The Promise of Access: Technology, Inequality, and the Political Economy of Hope* (Cambridge, MA: MIT Press, 2021). For a wide-ranging analysis of how geographic factors both shape and reflect the distribution of digital networks and the content that circulates through them, see Mark Graham and Martin Dittus, *Geographies of Digital Exclusion: Data and Inequality* (London: Pluto Press, 2022).

33. I develop an examination of the crucial role of IP geolocation in fostering geographic variegation in the digital economy in my article Luis F. Alvarez León, "The Digital Economy and Variegated Capitalism," *Canadian Journal of Communication* 40, no. 4 (2015): 637–54.

34. For a broader discussion into the historical development of GPS and how this spatial technology brought with it a new geo-epistemology, see William Rankin, *After the Map: Cartography, Navigation, and the Transformation of Territory in the Twentieth Century* (Chicago: University of Chicago Press, 2018).

35. The Global Positioning System was started by the US Department of Defense in 1978, with the first satellite launched in 1978 and the constellation of twenty-four satellites coming into operation in 1993. Today this system is operated by the United States Space Force. This is one of many Global Navigational Satellite Systems, which include: GLONASS, first launched in 1982 and achieving full coverage in 1994, begun as a Soviet system, and now owned and operated by Russia; BeiDou, a Chinese system that has undergone various iterations, first launched in 2000 and achieving global service by 2018; and the Galileo positioning system, a joint project by the European Union and the European Space Agency, which became operational in 2016 and launched its twenty-fourth and latest satellite in 2021. Other systems include regional, rather than global, navigation satellite systems such as NavIC, developed by the Indian Space Research Organization, and QZSS, also known as Michibiki, a Japanese system designed to augment the United States' GPS, in Japan and the Asia-Oceania region.

36. Paul E. Ceruzzi, *GPS* (Cambridge, MA: MIT Press, 2018), 115.

37. I draw this point from the work of Mark Granovetter on the economic dimensions of trust and its role in underpinning the social embeddedness of markets. See Mark Granovetter, *Society and Economy* (Cambridge, MA: Harvard University Press, 2017).

38. Ben Tarnoff documents this unprecedented process of privatization of the internet, its commercial turn, and its profound consequences for the construction of today's digital ecosystem. As he forcefully argues, this ecosystem is built on the profit motives of private firms, the enormous power of a handful

of very large corporations, and the narrowing spaces for non-commercial and civic-oriented spaces online. Like many other markets, this construction was not natural or spontaneous, but the result of deliberate state action that created the conditions for private interests to take over a space that had been built using public resources. See Ben Tarnoff, *Internet for the People: The Fight for Our Digital Future* (London: Verso, 2022).

39. For a legal analysis of this case, and how it relates to broader questions of jurisdiction, power, and control of the internet, see Jack Goldsmith and Tim Wu, *Who Controls the Internet? Illusions of a Borderless World* (Oxford: Oxford University Press, 2006).

40. This section specifically prohibits wearing or displaying uniforms, insignias, or emblems reminiscent of organizations or persons responsible for crimes against humanity: "*Est puni de l'amende prévue pour les contraventions de la 5e classe le fait, sauf pour les besoins d'un film, d'un spectacle ou d'une exposition comportant une évocation historique, de porter ou d'exhiber en public un uniforme, un insigne ou un emblème rappelant les uniformes, les insignes ou les emblèmes qui ont été portés ou exhibés soit par les membres d'une organisation déclarée criminelle en application de l'article 9 du statut du tribunal militaire international annexé à l'accord de Londres du 8 août 1945, soit par une personne reconnue coupable par une juridiction française ou internationale d'un ou plusieurs crimes contre l'humanité prévus par les articles 211-1 à 212-3 ou mentionnés par la loi no. 64-1326 du 26 décembre 1964.*" Légifrance, "Code Pénal, Article R645-1," accessed August 7, 2022, https://www.legifrance.gouv.fr /codes/article_lc/LEGIARTI000022375941.

41. Superior Court of Paris, "LICRA and UEJF vs Yahoo! Inc. and Yahoo France," May 22, 2000, http://www.lapres.net/yahen.html.

42. Yaman Akdeniz, "Case Analysis of *League Against Racism and Antisemitism (LICRA), French Union of Jewish Students, v. Yahoo! Inc. (USA), Yahoo France*, Tribunal de Grande Instance de Paris (The County Court of Paris), Interim Court Order, 20 November, 2000," *Electronic Business Law Reports* 1 (2001): 2.

43. Ben Lutkevich and John Burke, "What Is DNS? How Domain Name System Works," TechTarget Network Infrastructure, August 2021, https://www .techtarget.com/searchnetworking/definition/domain-name-system.

44. Cyril Houri, Method and systems for locating geographical locations of online users, United States US6665715B1, filed April 3, 2000, and issued December 16, 2003, https://patents.google.com/patent/US6665715B1/en?assignee=cyril +houri&oq=cyril+houri&sort=old.

45. Goldsmith and Wu, *Who Controls the Internet?*

46. Elsewhere I have developed a more thorough elaboration of some of these arguments linking the creation of digital markets, the politics and policies associated with them, and the role of geographic information (and geolocation in particular) in this process. See the following selection of journal articles and chapters: Cameran Ashraf and Luis F. Alvarez Leon, "The Logics and Territorialities of Geoblocking," in *Geoblocking and Global Video Culture*, ed. Ramon Lobato and James Meese (Amsterdam: Institute of Networked Cultures, 2016), 42–53; Luis F. Alvarez León, "The Digital Economy and Variegated

Capitalism," *Canadian Journal of Communication* 40, no. 4 (2015): 637–54; Luis F. Alvarez León, "The Political Economy of Spatial Data Infrastructures," *International Journal of Cartography* 4, no. 2 (2018): 151–69, https://doi.org /10.1080/23729333.2017.1371475; Alvarez León, "Information Policy and the Spatial Constitution of Digital Geographic Information Markets," *Economic Geography* 94, no. 3 (2018): 217–37. https://doi.org/10.1080/00130095.2017 .1388161.

CHAPTER 4. EYES IN THE SKY AND THE DIGITAL PLANET

1. Straits Research, "Satellite Data Services Market Size, Share and Analysis 2031," Satellite Data Services Market, 2023, https://straitsresearch.com/report /satellite-data-services-market.

2. Satellite Imaging Corporation, "Planetscope—Dove Satellite Constellation (3m)," 2022, https://www.satimagingcorp.com/satellite-sensors/other-satellite -sensors/dove-3m/.

3. Eric Berger, "SpaceX Launches 143 Satellites into Orbit, Most Ever [Updated]," *Ars Technica*, January 24, 2021, https://arstechnica.com/science/2021 /01/spacex-to-set-record-for-most-satellites-launched-on-a-single-mission/; Jonathan Amos, "SpaceX: World Record Number of Satellites Launched," BBC News, January 24, 2021, sec. Science & Environment, https://www.bbc.com /news/science-environment-55775977; Stephen Clark, "SpaceX Passes 2,500 Satellites Launched for Starlink Internet Network—Spaceflight Now," May 13, 2022, https://spaceflightnow.com/2022/05/13/spacex-passes-2500-satellites -launched-for-companys-starlink-network/.

4. Indian Space Research Organization and Antrix Corporation Limited, "PSLV-C37 Brochure" (ISRO, 2017), 4, https://www.isro.gov.in/pslv-c37-cartosat -2-series-satellite/pslv-c37-brochure-0.

5. Indian Space Research Organization and Antrix Corporation Limited, 5.

6. Parts of the discussion of these three developments and other ideas in this chapter are adapted from Luis F. Alvarez León, "An Emerging Satellite Ecosystem and the Changing Political Economy of Remote Sensing," in *The Nature of Data: Infrastructures, Environments, Politics*, ed. Jenny Goldstein and Eric Nost (Lincoln: University of Nebraska Press, 2022), 71–102.

7. For a thorough examination of the technological, political, and epistemic implications of the widespread use of sensors across domains and environments, see Jennifer Gabrys, *Program Earth* (Minneapolis: University of Minnesota Press, 2016).

8. This argument and some of the examples of environmental monitoring technologies are from Karen Bakker and Max Ritts, "Smart Earth: Environmental Governance in a Wired World," in *The Nature of Data: Infrastructures, Environments, Politics*, ed. Jenny Goldstein and Eric Nost (Lincoln: University of Nebraska Press, 2022), 74.

9. Cindy Lin, "How Forest Became Data: The Remaking of Ground Truth in Indonesia," in *The Nature of Data: Infrastructures, Environments, Politics*, ed. Jenny Goldstein and Eric Nost (Lincoln: University of Nebraska Press, 2022), 299.

10. As of May 14, 2021, the base has been renamed Vandenberg Space Force Base following the redesignation of the 30th Space Wing as Space Launch Delta 30 under the recently created US Space Force. Oscar Flores, "Vandenberg AFB Renames Base and 30th Space Wing," NBC Los Angeles, *Space Force* (blog), May 12, 2021, https://www.nbclosangeles.com/news/local/vandenberg-afb -renames-base-and-30th-space-wing/2593887/.

11. "Seasat—Earth Missions," NASA Jet Propulsion Laboratory (JPL), California Institute of Technology, accessed August 25, 2022, https://www.jpl.nasa .gov/missions/seasat.

12. "Trailblazer Sea Satellite Marks Its Coral Anniversary," NASA Jet Propulsion Laboratory (JPL), California Institute of Technology, June 27, 2013, https:// www.jpl.nasa.gov/news/trailblazer-sea-satellite-marks-its-coral-anniversary.

13. Molly Rettig, "Old Satellite Imagery Offers New Baseline Data," *Anchorage Daily News*, June 22, 2013, https://archive.ph/7PqBy.

14. Julia Rosen, "Shifting Ground: Fleets of Radar Satellites Are Measuring Movements on Earth Like Never Before," *Science*, February 25, 2021, https:// doi.org/10.1126/science.abh2435.

15. "A Revolution in Synthetic Aperture Radar (SAR) Data Earth Observation," ICEYE, 2022, 5, https://www.iceye.com/hubfs/Downloadables/SAR _Data_Brochure_ICEYE.pdf.

16. Rosen, "Shifting Ground."

17. The CORONA program was a joint effort by the US Department of Defense and the Central Intelligence Agency, operating from 1960 to 1972. CORONA was motivated by the successful launch of Sputnik by the USSR in October 1957, which led President Dwight Eisenhower to endorse the program in 1958. Further urgency for the operational capabilities of a satellite reconnaissance program came in the aftermath of the incident involving Gary Powers, who was captured by the Soviet government after his U-2 spy plane was shot down Near Aramil, Sverdlovsk Oblast, Soviet Union, in May 1960, during a photographic aerial reconnaissance mission. The CORONA program, which largely came to substitute the U-2 plane's photo reconnaissance functions, remained classified until February 1995. The official information center with archives, documents, and press releases documenting the program can be found at National Reconnaissance Office, "The CORONA Program," accessed August 27, 2022, https://www.nro.gov/History-and-Studies/Center-for-the-Study -of-National-Reconnaissance/The-CORONA-Program/. For declassified and digitized imagery from the CORONA program, as well as the LANYARD and ARGON satellite systems see United States Geological Survey's Earth Resource Observation and Science (EROS) Center, "Products Overview," *USGS EROS Archive* (blog), July 19, 2019, https://www.usgs.gov/centers/eros/science/usgs -eros-archive-products-overview.

18. For a history of the technological and institutional development that led to the creation and long-running operation of the Landsat satellite system, see Pamela Etter Mack, *Viewing the Earth: The Social Construction of the Landsat Satellite System*, Inside Technology (Cambridge, MA: MIT Press, 1990).

19. For an examination of the inherent difficulty of achieving a processual understanding of human and environmental dynamics through the use of

satellite remote sensing in isolation, see Matthew D. Turner and Peter J. Taylor, "Critical Reflections on the Use of Remote Sensing and GIS Technologies in Human Ecological Research," *Human Ecology* 31, no. 2 (June 2003): 177–82, https://doi.org/10.1023/A:1023958712140. For a more general examination of the relationship between pattern and process as a fundamental geographic question, see David O'Sullivan and George L. W. Perry, "Pattern, Process and Scale," in *Spatial Simulation: Exploring Pattern and Process* (Chichester, UK: John Wiley & Sons, Ltd, 2013), 29–56.

20. For an expansion on these arguments, as well as recommendations to rethink remote sensing concepts and practices, see Mia M. Bennett et al., "The Politics of Pixels: A Review and Agenda for Critical Remote Sensing," *Progress in Human Geography* 46, no. 3 (2022): 729–52, https://doi.org/10.1177/03091325221074691.

21. "Company," Planet Labs PBC, 2022, https://www.planet.com/company/.

22. Elspeth Lewis, "How Sputnik Changed the World," National Space Centre, March 10, 2017, https://spacecentre.co.uk/blog-post/sputnik-changed-world/.

23. Mark Wade, "R-7," *Encyclopedia Astronautica*, September 4, 2003, https://web.archive.org/web/20030904120332/http://www.astronautix.com/lvs/r7.htm.

24. Vincent L. Pisacane, "The Legacy of Transit: A Dedication," *Johns Hopkins Apl Technical Digest* 19, no. 1 (1998): 5–10. For more on the development of TRANSIT in the context of other satellite-enabled navigational and positioning systems, such as GPS, see Paul E. Ceruzzi, *GPS* (Cambridge, MA: MIT Press, 2018).

25. The personnel dossiers are stored by the National Archives and can be consulted at "Records of the Secretary of Defense (RG 330)," National Archives, August 15, 2016, https://www.archives.gov/iwg/declassified-records/rg-330-defense-secretary. For an overview of Operation Paperclip, see Annie Jacobsen, *Operation Paperclip: The Secret Intelligence Program That Brought Nazi Scientists to America* (New York: Little Brown and Company, 2015).

26. The list of NASA Honor Awards between 1969 and 1978 is in "SP-4012 NASA Historical Data Book: Volume IV. NASA Resources 1969–1978" (NASA), accessed August 28, 2022, https://history.nasa.gov/SP-4012/vol4/appa.htm. The latest release of the NASA Awards Historical Recipient List (V. 4.0 1959–2015) can be consulted using the NSSC Public Search engine at NASA, "Agency Awards Historical Recipient List" (NASA, n.d.), https://searchpub.nssc.nasa.gov/servlet/sm.web.Fetch?rhid=1000&did=2120817&type=released. Eventually, one of those awardees, Arthur Rudolph was investigated in 1984 by the US Government for war crimes, after which he renounced his US citizenship and left the country for West Germany in exchange for not being prosecuted by the Justice Department; he died in Hamburg in 1996, without ever having stood trial. See Wolfgang Saxon, "Arthur Rudolph, 89, Developer of Rocket in First Apollo Flight," *New York Times*, January 3, 1996, sec. U.S., https://www.nytimes.com/1996/01/03/us/arthur-rudolph-89-developer-of-rocket-in-first-apollo-flight.html.

27. For an examination of the importance of rocketry, and particularly the development of the V-2, for the Third Reich, see Michael J. Neufeld, *The Rocket*

and the Reich: Peenemünde and the Coming of the Ballistic Missile Era (Washington DC: Smithsonian Institution, 2013); Michael B. Petersen, *Missiles for the Fatherland: Peenemünde, National Socialism, and the V-2 Missile*, Cambridge Centennial of Flight (Cambridge: Cambridge University Press, 2011). On the use of slave labor for the construction of the Nazi rocketry program and the V-2 specifically, with a focus on documenting the Dora camp, see Andre Sellier, *A History of the Dora Camp: The Untold Story of the Nazi Slave Labor Camp That Secretly Manufactured V-2 Rockets* (Chicago: Ivan R. Dee, 2003). For an overview of the US space program and its culmination with the July 20,1969 Moon landing, the various developmental paths of rocketry, and the instrumental role played by von Braun and other Nazi scientists relocated by Operation Paperclip in its development, see Douglas Brinkley, *American Moonshot: John F. Kennedy and the Great Space Race* (New York: HarperCollins, 2019).

28. Richard Hollingham, "V2: The Nazi Rocket That Launched the Space Age," BBC, September 7, 2014, https://www.bbc.com/future/article/20140905 -the-nazis-space-age-rocket.

29. Kathryn Hambleton, "Around the Moon with NASA's First Launch of SLS with Orion," NASA, March 7, 2018, http://www.nasa.gov/feature/around -the-moon-with-nasa-s-first-launch-of-sls-with-orion.

30. "Artemis," NASA, accessed August 29, 2022, https://www.nasa.gov /specials/artemis/index.html.

31. "Artemis."

32. "Artemis."

33. "Artemis."

34. "Value of NASA," NASA, accessed July 20, 2023, https://www.nasa.gov /specials/value-of-nasa/index.html.

35. Kathryn Hambleton, "Artemis I Overview," NASA, February 20, 2018, http://www.nasa.gov/content/artemis-i-overview.

36. Amanda Kooser, "Elon Musk Breaks Down the Starship Numbers for a Million-Person SpaceX Mars Colony," CNET, January 16, 2020, https://www .cnet.com/science/elon-musk-drops-details-for-spacexs-million-person-mars -mega-colony/.

37. Mark Stokes et al., "China's Space and Counterspace Capabilities and Activities," US-China Economic and Security Review Commission, March 30, 2020, 86, https://www.uscc.gov/research/chinas-space-and-counterspace-activities.

38. China Great Wall Industry Corporation (CGWIC), "Satellite Communication, " accessed August 28, 2022, http://www.cgwic.com/Communications Satellite/project.html.

39. Rui C. Barbosa, "China Launch VENESAT-1—Debut Bird for Venezuela," *NASA Space Flight (NSF)*, October 29, 2008, https://www.nasaspaceflight .com/2008/10/china-launch-venesat/.

40. "Venesat-1 queda fuera de servicio," Latam Satelital, March 29, 2020, http://latamsatelital.com/venesat-1-queda-fuera-de-servicio/.

41. Barbosa, "China Launch VENESAT-1."

42. Stephen Clark, "China Launches Earth-Observing Satellite for Venezuela," Space.com, October 1, 2012, https://www.space.com/17849-china-satellite -launch-venezuela.html.

43. Stephen Clark, "China Successfully Launches Earth-Imaging Satellite for Venezuela," *Space Flight Now* (blog), October 9, 2017, https://spaceflightnow.com /2017/10/09/china-successfully-launches-earth-imaging-satellite-for-venezuela/.

44. Clark, "China Launches Earth-Observing Satellite for Venezuela."

45. M, "The Space Center Kenya Doesn't Own," Owaahh, March 31, 2016, https://owaahh.com/space-center-kenya-doesnt/; Joseph Ibeh, "Kenya and Italy Close in on Signing Ownership Deal in Respect of Luigi Broglio Space Centre," *Space in Africa* (blog), July 19, 2019, https://africanews.space/ kenya-and-italy-signing-ownership-deal-luigi-broglio-space-centre/.

46. "Europe's Spaceport," European Space Agency, accessed July 20, 2023, https://www.esa.int/Enabling_Support/Space_Transportation/Europe_s _Spaceport/Europe_s_Spaceport2. For more on the colonial relationships that shaped the development of the spaceport in French Guiana, as well as previous projects integral to the French state, such as Devil's Island penal colony in the island of Cayenne, see Peter Redfield, *Space in the Tropics: From Convicts to Rockets in French Guiana* (Berkeley: University of California Press, 2000).

47. "SpaceTech Industry 2021 / Q2 Landscape Overview," SpaceTech Analytics, May 2021), 9–10.

48. "SpaceTech Industry 2021," 21.

49. Elizabeth Mabrouk, "What Are SmallSats and CubeSats?," NASA, March 13, 2015, http://www.nasa.gov/content/what-are-smallsats-and-cubesats.

50. For an extended argument and analysis of how information policy regimes are crucial for the geographic constitution of information markets, see my article Luis F. Alvarez León, "Information Policy Regimes and the Spatial Constitution of Digital Geographic Information Markets," *Economic Geography* 94, no. 3 (2018): 217–37.

51. For an in-depth history of the institutional dimensions behind the creation and shaping of the Landsat program during its first two decades, see Mack, *Viewing the Earth*.

52. William Emery and Adriano Camps, "The History of Satellite Remote Sensing," in *Introduction to Satellite Remote Sensing* (Amsterdam: Elsevier, 2017), 26.

53. Zbigniew Brzezinski, "Presidential Directive NSC-54: Civil Operational Remote Sensing," The White House, November 16, 1979, 1–2, https://www .jimmycarterlibrary.gov/assets/documents/PD_54.pdf.

54. Brzezinski, 2.

55. Emery and Camps, "The History of Satellite Remote Sensing," 26.

56. Emery and Camps, 26.

57. Vincent Del Giudice, "The government, saying it ran out of money, pulled . . . ," UPI Archives, March 3, 1989, https://www.upi.com/Archives/1989 /03/02/The-government-saying-it-ran-out-of-money-pulled/6994604818000/.

58. Del Giudice.

59. John Noble Wilford, "US Halts Plan to Turn Off the Landsat Satellites," *New York Times*, March 17, 1989.

60. Anusuya Datta, "Landsat Completes 45 Years: Tracing the Journey," *Geospatial World* (blog), July 24, 2017, https://www.geospatialworld.net/blogs /landsat-completes-45-years/.

61. United States Congress, "Land Remote Sensing Policy Act of 1992," Pub. L. No. 102-555 (1992), 1, https://www.govinfo.gov/app/details/https%3A%2F%2Fwww.govinfo.gov%2Fapp%2Fdetails%2FCOMPS-1849.

62. "Revisiting the Land Remote Sensing Policy Act of 1992" National Geospatial Advisory Committee Landsat Advisory Group (NGAC), April 2021, 6, https://www.fgdc.gov/ngac/meetings/april-2021/ngac-paper-revisiting-the-land -remote-sensing.pdf.

63. *Wired* staff, "Clinton Unscrambles GPS Signals," *Wired*, May 1, 2000, https://www.wired.com/2000/05/clinton-unscrambles-gps-signals/.

CHAPTER 5. PEOPLE, PLATFORMS, AND ROBOTS ON THE MOVE

1. For an in-depth discussion of how the incorporation of digital platforms is reshaping political, technological, and social relations, particularly in cities, leading to the emergence of an identifiable form of "platform urbanism," see Sarah Barns, *Platform Urbanism: Negotiating Platform Ecosystems in Connected Cities* (Singapore: Springer Nature, 2020).

2. "Global Ride Hailing Market Report 2022 Featuring Uber, Grab, ANI Technologies, Gett, Lyft, DiDi Chuxing, Delphi Automotive, Daimler BlaBlaCar and Didi Chuxing Technology—ResearchAndMarkets.Com," May 19, 2022, citing *Ride-Hailing Global Market Report 2022: End-User, by Service Type*, Research and Markets Report, https://www.businesswire.com/news/home/2022 0519005545/en/Global-Ride-Hailing-Market-Report-2022-Featuring-Uber -Grab-ANI-Technologies-Gett-Lyft-DiDi-Chuxing-Delphi-Automotive-Daimler -BlaBlaCar-and-Didi-Chuxing-Technology---ResearchAndMarkets.com.

3. For a thorough account of these promises and how they structure Uber's business model, along with the experiences of drivers and passengers, see the ethnographic study Alex Rosenblat, *Uberland: How Algorithms Are Rewriting the Rules of Work* (Oakland: University of California Press, 2018).

4. See, for example, the following critiques of Uber and other TNCs' practices in cities, with attention to their platforms' impacts on labor conditions, traffic, pollution, urban governance, and other key aspects of city life: Mi Diao, Hui Kong, and Jinhua Zhao, "Impacts of Transportation Network Companies on Urban Mobility," *Nature Sustainability* 4, no. 6 (June 2021): 494–500, https://doi .org/10.1038/s41893-020-00678-z; Katie J. Wells, Kafui Attoh, and Declan Cullen, "'Just-in-Place' Labor: Driver Organizing in the Uber Workplace," *Environment and Planning A: Economy and Space* 53, no. 2 (2021): 315–31, https://doi .org/10.1177/0308518X20949266; Kafui Attoh, Katie Wells, and Declan Cullen, "'We're Building Their Data': Labor, Alienation, and Idiocy in the Smart City," *Environment and Planning D: Society and Space* 37, no. 6 (2019): 1007–24, https://doi.org/10.1177/0263775819856626; Ruth Berins Collier, Veena B. Dubal, and Christopher Carter, "Disrupting Regulation, Regulating Disruption: The Politics of Uber in the United States," SSRN Scholarly Paper (Rochester, NY: Social Science Research Network, March 22, 2018), https://papers.ssrn.com/abstract =3147296; Josh Dzieza, "Revolt of the Delivery Workers," *The Verge*, September 13, 2021, https://www.theverge.com/22667600/delivery-workers-seamless

-uber-relay-new-york-electric-bikes-apps. Furthermore, in summer 2022, many of these critiques were substantiated with the release of the "Uber files," a cache of internal documents including 83,000 emails, iMessages, and WhatsApp messages between 2013 and 2017, which were made available to journalists by Mark MacGann, Uber's former chief lobbyist for Europe. For an overview of the Uber files and how they constitute evidence for many long-standing criticisms against the company, especially as they pertain to its propensity for systematically breaking local regulations in cities around the world, see the series "The Uber Files," *Guardian*, July 2022, https://www.theguardian.com/news/series/uber-files.

5. For an overview of the development of autonomous vehicles, see Hod Lipson and Melba Kurman, *Driverless: Intelligent Cars and the Road Ahead* (Cambridge MA: MIT Press, 2016). On a more critical register, Paris Marx has persuasively argued that the hype and hope behind AVs are part of a broader wave of technologically oriented solutionism to what should be better understood as collective problems of transportation, and that this reframing constitutes the enactment of Big Tech's vision of the future, as well as an expression of its preferred business models. Marx develops this argument examining multiple ongoing transportation technological "disruptions," from AVs to ride hailing, on-demand scooters, and Elon Musk's various tunnelling proposals. See Paris Marx, *Road to Nowhere: What Silicon Valley Gets Wrong about the Future of Transportation* (London: Verso, 2022).

6. This argument is developed in greater depth in my article with Yuko Aoyama: Luis F. Alvarez León and Yuko Aoyama, "Industry Emergence and Market Capture: The Rise of Autonomous Vehicles," *Technological Forecasting and Social Change* 180 (July 2022): 121661, https://doi.org/10.1016/j.techfore.2022.121661.

7. For a review of the capabilities and technical performance of the various sensors used by autonomous vehicles, their software and hardware aspects, and methods for sensor fusion, see De Jong Yeong et al., "Sensor and Sensor Fusion Technology in Autonomous Vehicles: A Review," *Sensors* 21, no. 6 (January 2021): 2140, https://doi.org/10.3390/s21062140.

8. According to the Society of Automobile Engineer's standard reference, as of 2021, the levels of driving automation are:

Level 0: No Driving Automation
Level 1: Driver Assistance
Level 2: Partial Driving Automation
Level 3: Conditional Driving Automation
Level 4: High Driving Automation
Level 5: Full Driving Automation

For more details on each of these levels, see On-Road Automated Driving (ORAD) Committee, "Taxonomy and Definitions for Terms Related to Driving Automation Systems for On-Road Motor Vehicles," SAE International, April 30, 2021, https://doi.org/10.4271/J3016_202104.

9. This is a point of contention because one of the arguments of AV proponents is that to produce safety gains these vehicles do not necessarily have to

drive flawlessly, but only marginally better than humans. What this argument misses, however, is that for AVs to be adopted, they must be first accepted by large sections of the public. Yet it may be difficult for the public to accept a transition to automated driving if it does not deliver drastically improved safety levels, since this may contravene many people's morals or the societal ethics. For an examination of the competing perspectives shaping the debate about which system of values, rules, or principles (e.g., societal ethics vs. individual morals) should govern the decisions made by artificial intelligence in autonomous vehicles, see E. Kassens-Noor, Josh Siegel, and Travis Decaminada, "Choosing Ethics over Morals: A Possible Determinant to Embracing Artificial Intelligence in Future Urban Mobility," *Frontiers in Sustainable Cities* 28, no. 3 (2021), https://www.frontiersin.org/articles/10.3389/frsc.2021.723475.

10. "Seaborne Trade—Capacity of Container Ships 2021," Statista, November 2021, https://www.statista.com/statistics/267603/capacity-of-container-ships-in-the-global-seaborne-trade/.

11. *The Future of Rail: Opportunities for Energy and the Environment,* International Energy Agency and International Union of Railways (IEA), 2019, 33, https://doi.org/10.1787/9789264312821-en.

12. Aviation data taken from reports issued in 2019 and 2022 by the International Civil Aviation Organization: "Economic Impact of COVID-19 on Civil Aviation," ICAO, June 10, 2022, https://www.icao.int/sustainability/Documents/Covid-19/ICAO_coronavirus_Econ_Impact.pdf; and "The World of Air Transport in 2019," International Civil Aviation Organization, https://www.icao.int/annual-report-2019/Pages/the-world-of-air-transport-in-2019.aspx.

13. Asia, Oceania, and the Middle East are counted as a single region under OICA's classification.

14. The question of Henry Ford's decision to increase wages dramatically in 1914 has led to a number of debates about its causes (see Daniel M. G. Raff and Lawrence H. Summers, "Did Henry Ford Pay Efficiency Wages?," *Journal of Labor Economics* 5, no. 4 [October 1987]: s57–86); its impacts on gender dynamics and family life (see Martha May, "The Historical Problem of the Family Wage: The Ford Motor Company and the Five Dollar Day," *Feminist Studies* 8, no. 2 [1982]: 399–424, https://doi.org/10.2307/3177569); and its broader consequences for social organization and labor-management relations (see Stephen Mayer III, *The Five Dollar Day: Labor Management and Social Control in the Ford Motor Company, 1908–1921* [Albany: State University of New York Press, 1981]).

15. Mimi Sheller and John Urry, "The City and the Car," *International Journal of Urban and Regional Research* 24, no. 4 (December 2000): 737–57, https://doi.org/10.1111/1468-2427.00276.

16. John Urry, "The 'System' of Automobility," *Theory, Culture & Society* 21, nos. 4–5 (2004): 25–26, https://doi.org/10.1177/0263276404046059. Italics in the original text.

17. Hannah Ritchie, "Cars, Planes, Trains: Where Do CO_2 Emissions from Transport Come From?," Our World in Data, October 6, 2020, https://ourworldindata.org/co2-emissions-from-transport.

18. Robert D. Bullard, Glenn S. Johnson, and Angel O. Torres, eds., *Highway Robbery: Transportation Racism and New Routes to Equity* (Cambridge, MA: South End Press, 2004), 4. For a legal and historical perspective on the role of transportation policy in driving racial inequality in the United States, see Deborah N. Archer, "Transportation Policy and the Underdevelopment of Black Communities," *Iowa Law Review* 106 (July 2021): 2125–51.

19. For instance, Robert L. French has documented the long historical development of vehicular navigation; see Robert L. French, "From Chinese Chariots to Smart Cars: 2,000 Years of Vehicular Navigation," *Navigation* 42, no. 1 (1995): 235–58, https://doi.org/10.1002/j.2161-4296.1995.tb02336.x.

20. For an assessment of the causes and consequences of this technological development in automobiles, see M. C. Forelle, "The Material Consequences of 'Chipification': The Case of Software-Embedded Cars," *Big Data and Society* 9, no. 1 (January 1, 2022): 20539517221095428, https://doi.org/10.1177/205395 17221095429.

21. For more on the consequences of the transformation of automobiles into mobile spatial media, and the role of on-board navigation systems in this process, see my article Luis F. Alvarez León, "How Cars Became Mobile Spatial Media: A Geographical Political Economy of On-Board Navigation," *Mobile Media and Communication* 7, no. 3 (2019): 362–79, https://doi.org/10.1177 /2050157919826356.

22. Marc Steinberg cautions us to avoid an automatic association between the idea of platforms and digital technologies. Further, he convincingly argues for a historicization of platforms that highlights the crucial role played by the automobile industry in their development throughout the twentieth century. This perspective, in turn, can help us gain better insight into how "platforms themselves are now completing their loops from automobile factories to smartphones back to the automobile with Uber, Didi, Tesla, Waymo, Apple and their experiments with autonomous driving and city mapping (Chen and Qiu, 2019), not to mention persistent rumours of tech companies getting into automobile production." Marc Steinberg, "From Automobile Capitalism to Platform Capitalism: Toyotism as a Prehistory of Digital Platforms," *Organization Studies* 43, no. 7 (2022): 1069-1090, https://doi.org/10.1177/0170840621103068.1.

23. Erum Salam, "'They Stole from Us': The New York Taxi Drivers Mired in Debt over Medallions," *Guardian*, October 2, 2021, sec. US news, https://www .theguardian.com/us-news/2021/oct/02/new-york-city-taxi-medallion-drivers -debt.

24. Sujeet Indap, "Wall Street Eyes NYC Taxis as Beleaguered Drivers Win Relief," *Financial Times*, November 15, 2021.

25. Brian M. Rosenthal, "N.Y.C. Cabbies Win Millions More in Aid after Hunger Strike," *New York Times*, November 3, 2021, sec. New York, https:// www.nytimes.com/2021/11/03/nyregion/nyc-taxi-drivers-hunger-strike.html.

26. Indap, "Wall Street Eyes NYC Taxis as Beleaguered Drivers Win Relief."

27. For a recent account of how Uber developed this urban disruption playbook through its operations in Washington, D.C., as well as the far-reaching effects of this disruption for urban politics, labor, transportation, and city life,

see Katie J. Wells, Kafui Ablode Attoh, and Declan Cullen, *Disrupting D.C.: The Rise of Uber and the Fall of the City* (Princeton: Princeton University Press, 2023).

28. The victim was a fifty-five-year-old businessman named Miguel Angel Lecuna, who was married to actress and TV presenter Georgina Barbarossa. Lecuna was ambushed and stabbed in the backseat of the taxi by two perpetrators who were allegedly acting in coordination with the driver. This crime set off a widespread debate in Argentine society about the "taxi mafia." See "Asesinan en un taxi en Palermo al marido de Georgina Barbarossa," *La Nación*, November 3, 2001, https://www.lanacion.com.ar/sociedad/asesinan-en-un-taxi-en -palermo-al-marido-de-georgina-barbarossa-nid348353/; Diario "Creen que la 'mafia de taxis' asesinó al marido de Georgina," *Diario El Dia de La Plata*, www.eldia.com, April 11, 2001), https://www.eldia.com/nota/2001-11-4-creen -que-la-mafia-de-taxis-asesino-al-marido-de-georgina.

29. Juan Manuel del Nido, *Taxis vs. Uber: Courts, Markets, and Technology in Buenos Aires* (Stanford, CA: Stanford University Press, 2021), 23, https://doi .org/10.1515/9781503629684.

30. On January 29, 2013, thirty-five-year-old taxi driver Rubén Darío Botta was speeding along Palermo avenue when his car impacted a motorcycle and crashed against a bus, which sent his vehicle against traffic along 9 de Julio avenue, where he hit three more cars and ran over six pedestrians, among them twenty-three-year-old Leonela Roble, who died instantly. See "El taxista que atropelló a Leonela no dijo que era diabético cuando sacó el registro," *La Nación*, February 26, 2013, https://www.lanacion.com.ar/seguridad/el-taxista-que-atropello -a-leonela-no-dijo-que-era-diabetico-cuando-saco-el-registro-nid1558234/; "La madre de Leonela pide justicia y rechaza la versión del taxista," *Clarín*, January 31, 2013, https://www.clarin.com/policiales/leonela-justicia-desconfia -hipotesis-glucemia_o_HygQWwssvXe.html.

31. Collier, et al., "Disrupting Regulation, Regulating Disruption, 4.

32. del Nido, *Taxis vs. Uber*, 25.

33. See Robert Booth, "Uber Drivers to Launch Legal Bid to Uncover App's Algorithm," *Guardian*, July 20, 2020, sec. Technology, https://www.theguardian .com/technology/2020/jul/20/uber-drivers-to-launch-legal-bid-to-uncover-apps -algorithm; Morgan Clendaniel, "Uber Says It Will Be More Transparent with Drivers about the Full Cost of the Ride," Fast Company, July 23, 2021, https:// www.fastcompany.com/90658669/uber-says-it-will-be-more-transparent-with -drivers-about-the-full-cost-of-the-ride; Sebastian Klovig Skelton, "Uber Drivers Strike over Pay Issues and Algorithmic Transparency, *ComputerWeekly*, July 22, 2022, https://www.computerweekly.com/news/252521871/Uber-drivers-strike -over-pay-issues-and-algorithmic-transparency. For a discussion that incorporates these information asymmetries into an analysis of Uber's operations, particularly with a focus on labor practices, see Rosenblat, *Uberland*.

34. del Nido, *Taxis vs. Uber*, 150–51.

35. del Nido, 151.

36. "Uber Cities—Rides Around the World," Uber, 2022, https://www.uber .com/global/en/cities/.

37. Jovanna Rosen and I develop this argument at length in Luis F. Alvarez León and Jovanna Rosen, "Technology as Ideology in Urban Governance," *Annals of the American Association of Geographers* 110, no. 2 (2020): 497–506, https://doi.org/10.1080/24694452.2019.1660139.

38. Wells et al., "'Just-in-Place' Labor," 317.

39. Wells et al., 325.

40. Mimi Sheller, *Mobility Justice: The Politics of Movement in the Age of Extremes* (London; Brooklyn, NY: Verso, 2018), 87.

CONCLUSION

1. Vincent Mosco, *The Digital Sublime: Myth, Power, and Cyberspace*, 1 (Cambridge, MA: MIT Press, 2005).

2. For some key contributions to this conversation, see Meredith Broussard, *Artificial Unintelligence: How Computers Misunderstand the World* (Cambridge, MA: MIT Press, 2018); Ruha Benjamin, *Race after Technology: Abolitionist Tools for the New Jim Code* (Medford, MA: Polity, 2019); Marie Hicks, *Programmed Inequality* (Cambridge, MA: MIT Press, 2017); Brian Jordan Jefferson, *Digitize and Punish: Racial Criminalization in the Digital Age* (Minneapolis: University of Minnesota Press, 2020); Jathan Sadowski, *Too Smart: How Digital Capitalism Is Extracting Data, Controlling Our Lives, and Taking over the World* (Cambridge, MA: MIT Press, 2020); Safiya Umoja Noble, *Algorithms of Oppression: How Search Engines Reinforce Racism* (New York: New York University Press, 2018).

3. To name a few: Tung-Hui Hu, *A Prehistory of the Cloud* (Cambridge, MA: MIT Press, 2015); Nicole Starosielski, *The Undersea Network* (Durham, NC: Duke University Press, 2015); Andrew Blum, *Tubes: A Journey to the Center of the Internet* (New York: HarperCollins, 2012); Ingrid Burrington, *Networks of New York: An Illustrated Field Guide to Urban Internet Infrastructure* (Brooklyn: Melville House, 2016); John P. Jacob, Luke Skrebowski, and Smithsonian American Art Museum, eds., *Trevor Paglen: Sites Unseen* (Washington, DC; London: Smithsonian American Art Museum in association with D. Giles Limited, 2018); Louise Amoore, *Cloud Ethics: Algorithms and the Attributes of Ourselves and Others* (Durham, NC: Duke University Press, 2020); Matthew A. Zook, *The Geography of the Internet Industry: Venture Capital, Dot-Coms, and Local Knowledge* (Malden, MA: Blackwell, 2005); Rob Kitchin and Martin Dodge, *Code/Space: Software and Everyday Life* (Cambridge, MA: MIT Press, 2011); Nick Lally, Kelly Kay, and Jim Thatcher, "Computational Parasites and Hydropower: A Political Ecology of Bitcoin Mining on the Columbia River," *Environment and Planning E: Nature and Space* 5, no. 1 (2019), https://doi.org/10.1177/2514848619867608; Daniel Greene, "Landlords of the Internet: Big Data and Big Real Estate," *Social Studies of Science* 52, no. 6 (2022): 904–27, https://doi.org/10.1177/03063127221124943.

4. For some historical examples of the digital networking efforts in other countries, like France, Chile, and the Soviet Union see, respectively, Eden Medina, *Cybernetic Revolutionaries: Technology and Politics in Allende's Chile*

(Cambridge, MA: MIT Press, 2014); Julien Mailland and Kevin Driscoll, *Minitel: Welcome to the Internet*, Platform Studies (Cambridge, MA: MIT Press, 2017); Benjamin Peters, *How Not to Network a Nation: The Uneasy History of the Soviet Internet* (Cambridge, MA: MIT Press, 2017).

5. Jovanna Rosen and Luis F. Alvarez León, "Signaling Hinterlands and the Spatial Networks of Digital Capitalism," *Annals of the American Association of Geographers* (2023): 1–13, https://doi.org/10.1080/24694452.2023.2249974; Jovanna Rosen and Luis F. Alvarez León, "The Digital Growth Machine: Urban Change and the Ideology of Technology," *Annals of the American Association of Geographers* 112, no. 8, (2022): 2248–65, https://doi.org/10.1080/24694452.2022.2052008.

6. For an account of the thoroughly emplaced nature of artificial intelligence and the industry that has developed around it, see Kate Crawford, *Atlas of AI: Power, Politics, and the Planetary Costs of Artificial Intelligence* (New Haven, CT: Yale University Press, 2021).

Bibliography

Abbate, Janet. *Inventing the Internet.* Cambridge, MA: MIT Press, 2000.

Abramitzky, Ran, Leah Platt Boustan, and Katherine Eriksson. "A Nation of Immigrants: Assimilation and Economic Outcomes in the Age of Mass Migration." *Journal of Political Economy* 122, no. 3 (2014): 467–506. https://doi.org/10.1086/675805.

Akdeniz, Yaman. "Case Analysis of *League Against Racism and Antisemitism (LICRA), French Union of Jewish Students, v Yahoo! Inc. (USA), Yahoo France*, Tribunal de Grande Instance de Paris (The County Court of Paris), Interim Court Order, 20 November 2000." *Electronic Business Law Reports* 1 (2001): 110–20.

Alvarez Leon, Luis Felipe. "Assembling Digital Economies: Geographic Information Markets and Intellectual Property Regimes in the United States and the European Union." PhD diss., UCLA, 2016.

Alvarez León, Luis F. "The Digital Economy and Variegated Capitalism." *Canadian Journal of Communication* 40, no. 4 (2015): 637–54

———. "The Emergence of Netflix and the New Digital Economic Geography of Hollywood." In *Netflix at the Nexus: Content, Practice, and Production in the Age of Streaming Television*, edited by Amber M. Buck and Theo Plothe, 47–63. New York: Peter Lang, 2019.

———. "An Emerging Satellite Ecosystem and the Changing Political Economy of Remote Sensing." In *The Nature of Data: Infrastructures, Environments, Politics*, edited by Jenny Goldstein and Eric Nost, 71–102. Lincoln: University of Nebraska Press, 2022.

———. "How Cars Became Mobile Spatial Media: A Geographical Political Economy of On-Board Navigation." *Mobile Media and Communication* 7, no. 3 (2019): 362–79. https://doi.org/10.1177/2050157919826356.

———. "Information Policy and the Spatial Constitution of Digital Geographic Information Markets." *Economic Geography* 94, no. 3 (2018): 217–37. https://doi.org/10.1080/00130095.2017.1388161.

———. "Information Policy Regimes and the Spatial Constitution of Digital Geographic Information Markets." *Economic Geography* 94, no. 3 (2018): 217–37.

———. "The Political Economy of Spatial Data Infrastructures." *International Journal of Cartography* 4, no. 2 (2018): 151–69. https://doi.org/10.1080/23729333.2017.1371475.

———. "Property Regimes and the Commodification of Geographic Information: An Examination of Google Street View." *Big Data & Society* 3, no. 2 (2016): 2053951716637885. https://doi.org/10.1177/2053951716637885.

Alvarez León, Luis F., and Yuko Aoyama. "Industry Emergence and Market Capture: The Rise of Autonomous Vehicles." *Technological Forecasting and Social Change* 180 (July 2022): 121661. https://doi.org/10.1016/j.techfore.2022.121661.

Alvarez León, Luis F., Brett Christophers, and Leqian Yu, eds. "The Spatial Constitution of Markets (Special Issue)." *Economic Geography* 94, no. 3 (2018).

Alvarez León, Luis F., and Jovanna Rosen. "Technology as Ideology in Urban Governance." *Annals of the American Association of Geographers* 110, no. 2 (2020): 497–506. https://doi.org/10.1080/24694452.2019.1660139.

Alvarez Leon, Luis, and Sterling Quinn. "Street-Level Imagery." *Geographic Information Science & Technology Body of Knowledge* 2019, no. Q4 (2019). https://doi.org/10.22224/gistbok/2019.4.12.

———. "The Value of Crowdsourced Street-Level Imagery: Examining the Shifting Property Regimes of OpenStreetCam and Mapillary." *GeoJournal* 84, no. 2 (April 2019): 395–414. https://doi.org/10.1007/s10708-018-9865-4.

Amoore, Louise. *Cloud Ethics: Algorithms and the Attributes of Ourselves and Others*. Durham, NC: Duke University Press, 2020.

Amos, Jonathan. "SpaceX: World Record Number of Satellites Launched." BBC News, January 24, 2021, sec. Science & Environment. https://www.bbc.com/news/science-environment-55775977.

Archer, Deborah N. "Transportation Policy and the Underdevelopment of Black Communities." *Iowa Law Review* 106 (July 2021): 2125–51.

Ashraf, Cameran, and Luis F Alvarez Leon. "The Logics and Territorialities of Geoblocking." In *Geoblocking and Global Video Culture*, edited by Ramon Lobato and James Meese, 42–53. Amsterdam: Institute of Networked Cultures, 2016.

Attoh, Kafui, Katie Wells, and Declan Cullen. "'We're Building Their Data': Labor, Alienation, and Idiocy in the Smart City." *Environment and Planning D: Society and Space* 37, no. 6 (2019): 1007–24. https://doi.org/10.1177/0263775819856626.

Bakker, Karen, and Max Ritts. "Smart Earth: Environmental Governance in a Wired World." In *The Nature of Data: Infrastructures, Environments,*

Politics, edited by Jenny Goldstein and Eric Nost, 61–82. Lincoln: University of Nebraska Press, 2022.

Balland, Pierre-Alexandre, Ron Boschma, Joan Crespo, and David L. Rigby. "Smart Specialization Policy in the European Union: Relatedness, Knowledge Complexity and Regional Diversification." *Regional Studies* 53, no. 9 (2019): 1252–68. https://doi.org/10.1080/00343404.2018.1437900.

Balland, Pierre-Alexandre, and David Rigby. "The Geography of Complex Knowledge." *Economic Geography* 93, no. 1 (January 1, 2017): 1–23. https://doi.org/10.1080/00130095.2016.1205947.

Baran, Paul. "On Distributed Communications: I. Introduction to Distributed Communications Networks." RAND Corporation, August 1964. https://www.rand.org/pubs/research_memoranda/RM3420.html.

Barbosa, Rui C. "China Launch VENESAT-1—Debut Bird for Venezuela." *NASASpaceFlight, NSF*, October 29, 2008. https://www.nasaspaceflight.com/2008/10/china-launch-venesat/.

Barlow, John Perry. "A Declaration of the Independence of Cyberspace." Electronic Frontier Foundation, 1996. https://www.eff.org/cyberspace-independence.

Barnes, Trevor J. "A Morality Tale of Two Location Theorists in Hitler's Germany: Walter Christaller and August Lösch." In *Hitler's Geographies: The Spatialities of the Third Reich*, edited by Paolo Giaccaria and Claudio Minca, 198–217. Chicago: University of Chicago Press, 2016.

———. "'Desk Killers': Walter Christaller, Central Place Theory, and the Nazis." In *Geographies of Knowledge and Power*, edited by Peter Meusburger, Derek Gregory, and Laura Suarsana, 187–201. Dordrecht: Springer, 2015.

———. "Reopke Lecture in Economic Geography: Notes from the Underground: Why the History of Economic Geography Matters: The Case of Central Place Theory." *Economic Geography* 88, no. 1 (2012): 1–26. https://doi.org/10.1111/j.1944-8287.2011.01140.x.

Barnes, Trevor J., and Brett Christophers. *Economic Geography: A Critical Introduction*. Critical Introductions to Geography. Hoboken, NJ: John Wiley & Sons, 2018.

Barnes, Trevor J., and Claudio Minca. "Nazi Spatial Theory: The Dark Geographies of Carl Schmitt and Walter Christaller." *Annals of the Association of American Geographers* 103 (2013): 669–87. https://doi.org/10.2307/23485411?ref=no-x-route:2ca919d517fd198be3b5a9b6c83ce18a.

Barns, Sarah. *Platform Urbanism: Negotiating Platform Ecosystems in Connected Cities*. Singapore: Springer Nature, 2020.

Bar-Zeev, Avi. "Was Google Earth Stolen? (No)." *Medium* (blog), October 7, 2022. https://avibarzeev.medium.com/was-google-earth-stolen-7d1b821e589b.

Bathelt, Harald, Anders Malmberg, and Peter Maskell. "Clusters and Knowledge: Local Buzz, Global Pipelines and the Process of Knowledge Creation." *Progress in Human Geography* 28, no. 1 (2004): 31–56. https://doi.org/10.1191/0309132504ph469oa.

Benjamin, Ruha. *Race after Technology: Abolitionist Tools for the New Jim Code*. Medford, MA: Polity, 2019.

Benkler, Yochai, Rob Faris, and Hal Roberts. *Network Propaganda: Manipulation, Disinformation, and Radicalization in American Politics*. New York: Oxford University Press, 2018.

Bennett, Mia M., Janice Kai Chen, Luis F. Alvarez Leon, and Colin J. Gleason. "The Politics of Pixels: A Review and Agenda for Critical Remote Sensing." *Progress in Human Geography* 46, no. 3 (2022): 729-752. https://doi.org/10.1177/03091325221074691.

Berger, Eric. "SpaceX Launches 143 Satellites into Orbit, Most Ever [Updated]." *Ars Technica*, January 24, 2021. https://arstechnica.com/science/2021/01/spacex-to-set-record-for-most-satellites-launched-on-a-single-mission/.

Bergmann, Luke, and Nick Lally. "For Geographical Imagination Systems." *Annals of the American Association of Geographers* 111, no. 1 (2021): 26–35. https://doi.org/10.1080/24694452.2020.1750941.

Berndt, Christian, and Marc Boeckler. "Geographies of Circulation and Exchange: Constructions of Markets." *Progress in Human Geography* 33, no. 4 (2009): 535–51. https://doi.org/10.1177/0309132509104805.

———. "Geographies of Marketization." In *The Wiley-Blackwell Companion to Economic Geography*, edited by Trevor Barnes, Jamie Peck, and Eric Sheppard, 199–212. Chichester, UK: John Wiley & Sons, Ltd, 2012. http://doi.wiley.com/10.1002/9781118384497.ch12.

Berndt, Christian, Jamie Peck, and Norma M. Rantisi, eds. *Market/Place: Exploring Spaces of Exchange*. Economic Transformations. Newcastle upon Tyne: Agenda Publishing, 2020.

Blum, Andrew. *Tubes: A Journey to the Center of the Internet*. New York: HarperCollins, 2012.

Booth, Robert. "Uber Drivers to Launch Legal Bid to Uncover App's Algorithm." *Guardian*, July 20, 2020, sec. Technology. https://www.theguardian.com/technology/2020/jul/20/uber-drivers-to-launch-legal-bid-to-uncover-apps-algorithm.

Boschma, Ron, Pierre-Alexandre Balland, and Dieter Franz Kogler. "Relatedness and Technological Change in Cities: The Rise and Fall of Technological Knowledge in US Metropolitan Areas from 1981 to 2010." *Industrial and Corporate Change* 24, no. 1 (February 1, 2015): 223–50. https://doi.org/10.1093/icc/dtu012.

boyd, danah, and Kate Crawford. "Critical Questions for Big Data." *Information, Communication and Society* 15, no. 5 (2012): 662–79. https://doi.org/10.1080/1369118X.2012.678878.

Brenner, Neil, Jamie Peck, and Nik Theodore. "After Neoliberalization?" *Globalizations* 7, no. 3 (2010) : 327–45. https://doi.org/10.1080/14747731003669669.

Brin, Sergey, and Lawrence Page. "The Anatomy of a Large-Scale Hypertextual Web Search Engine." *Computer Networks and ISDN Systems*, Proceedings of the Seventh International World Wide Web Conference, 30, no. 1 (April 1998): 107–17. https://doi.org/10.1016/S0169-7552(98)00110-X.

Brinkley, Douglas. *American Moonshot: John F. Kennedy and the Great Space Race*. New York City: HarperCollins, 2019.

Broussard, Meredith. *Artificial Unintelligence: How Computers Misunderstand the World*. Cambridge, MA: MIT Press, 2018.

Brzezinski, Zbigniew. "Presidential Directive NSC-54: Civil Operational Remote Sensing." The White House, November 16, 1979. https://www.jimmycarter library.gov/assets/documents/PD_54.pdf.

Bullard, Robert D., Glenn S. Johnson, and Angel O. Torres, eds. *Highway Robbery: Transportation Racism and New Routes to Equity*. Cambridge, MA: South End Press, 2004.

Burns, Ryan, Craig M. Dalton, and Jim E. Thatcher. "Critical Data, Critical Technology in Theory and Practice." *The Professional Geographer* 70, no. 1 (2018): 126–28. https://doi.org/10.1080/00330124.2017.1325749.

Burrington, Ingrid. *Networks of New York: An Illustrated Field Guide to Urban Internet Infrastructure*. Brooklyn: Melville House, 2016.

Business Wire. "Global Ride Hailing Market Report 2022 Featuring Uber, Grab, ANI Technologies, Gett, Lyft, DiDi Chuxing, Delphi Automotive, Daimler BlaBlaCar and Didi Chuxing Technology—ResearchAndMarkets.Com." May 19, 2022. https://www.businesswire.com/news/home/20220519005545 /en/Global-Ride-Hailing-Market-Report-2022-Featuring-Uber-Grab-ANI -Technologies-Gett-Lyft-DiDi-Chuxing-Delphi-Automotive-Daimler-BlaBla Car-and-Didi-Chuxing-Technology---ResearchAndMarkets.com.

Cairncross, Frances C. *The Death of Distance: How the Communications Revolution Is Changing Our Lives*. 2nd ed. Boston: Harvard Business Press, 1997.

Callejo, Patricia, Marco Gramaglia, Rubén Cuevas, and Ángel Cuevas. "A Deep Dive into the Accuracy of IP Geolocation Databases and Its Impact on Online Advertising." *IEEE Transactions on Mobile Computing* 22, no. 8 (2022): 1–15. https://doi.org/10.1109/TMC.2022.3166785.

Caquard, Sébastien. "Cartographies of Fictional Worlds: Conclusive Remarks." *Cartographic Journal* 48, no. 4 (2011): 224–25. https://doi.org/10.1179/000 870411X13203362557264.

Castells, Manuel. *The Internet Galaxy*. Oxford: Oxford University Press, 2001.

Ceruzzi, Paul E. *GPS*. Cambridge, MA: MIT Press, 2018.

Chakraborty, Abhishek, and Akshay Narayan Hargude. "Dabbawala: Introducing Technology to the Dabbawalas of Mumbai." In *Proceedings of the 17th International Conference on Human-Computer Interaction with Mobile Devices and Services Adjunct*, 660–67. Copenhagen Denmark: ACM, 2015. https://doi.org/10.1145/2786567.2793685.

China Great Wall Industry Corporation (CGWIC). "Satellite Communication." Accessed August 28, 2022. http://www.cgwic.com/CommunicationsSatellite /project.html.

Christophers, Brett. *Envisioning Media Power: On Capital and Geographies of Television*. Lanham, MD: Lexington Books, 2009.

———. "The Territorial Fix: Price, Power and Profit in the Geographies of Markets." *Progress in Human Geography* 38 (January 7, 2015): 754–70. https:// doi.org/10.1177/0309132513516176.

Christopherson, Susan, and Michael Storper. "The City as Studio, the World as Backlot: The Impact of Vertical Disintegration on the Location of the

Motion-Picture Industry." *Environment and Planning D: Society and Space* 4, no. 3 (1986): 305–20.

Clarín. "La madre de Leonela pide justicia y rechaza la versión del taxista." January 31, 2013. https://www.clarin.com/policiales/leonela-justicia-desconfia-hipotesis-glucemia_o_HygQWwssvXe.html.

Clark, Stephen. "China Launches Earth-Observing Satellite for Venezuela." Space.com, October 1, 2012. https://www.space.com/17849-china-satellite-launch-venezuela.html.

———. "China Successfully Launches Earth-Imaging Satellite for Venezuela." Spaceflight Now, October 9, 2017. https://spaceflightnow.com/2017/10/09/china-successfully-launches-earth-imaging-satellite-for-venezuela/.

———. "SpaceX Passes 2,500 Satellites Launched for Starlink Internet Network." Spaceflight Now, May 13, 2022. https://spaceflightnow.com/2022/05/13/spacex-passes-2500-satellites-launched-for-companys-starlink-network/.

Clendaniel, Morgan. "Uber Says It Will Be More Transparent with Drivers about the Full Cost of the Ride." Fast Company, July 23, 2021. https://www.fastcompany.com/90658669/uber-says-it-will-be-more-transparent-with-drivers-about-the-full-cost-of-the-ride.

Cohn, D'vera. "How U.S. Immigration Laws and Rules Have Changed through History." *Pew Research Center* (blog), September 2015. https://www.pewresearch.org/short-reads/2015/09/30/how-u-s-immigration-laws-and-rules-have-changed-through-history/.

Collier, Ruth Berins, Vena B. Dubal, and Christopher Carter. "Disrupting Regulation, Regulating Disruption: The Politics of Uber in the United States." SSRN Scholarly Paper. Rochester, NY: Social Science Research Network, March 22, 2018. https://papers.ssrn.com/abstract=3147296.

Connor, Dylan Shane. "The Cream of the Crop? Geography, Networks, and Irish Migrant Selection in the Age of Mass Migration." *Journal of Economic History* 79, no. 1 (2019): 139–75. https://doi.org/10.1017/S0022050718000682.

Crain, Matthew. *Profit over Privacy: How Surveillance Advertising Conquered the Internet*. Minneapolis: University of Minnesota Press, 2021.

Crampton, Jeremy W. "Collect It All: National Security, Big Data and Governance." *GeoJournal* 80 (2015): 519–31. https://doi.org/10.1007/s10708-014-9598-y.

———. "Keyhole, Google Earth, and 3D Worlds: An Interview with Avi Bar-Zeev." *Cartographica: The International Journal for Geographic Information and Geovisualization* 43, no. 2 (Summer 2008): 85–93. https://doi.org/10.3138/carto.43.2.85.

———. *Mapping: A Critical Introduction to Cartography and GIS*. Critical Introductions to Geography. Malden, MA: Wiley-Blackwell, 2010.

———. "Mappings." *The Wiley-Blackwell Companion to Cultural Geography*. Malden, MA: Wiley-Blackwell, 2013, 423–36.

Crampton, Jeremy W., Mark Graham, Ate Poorthuis, Taylor Shelton, Monica Stephens, Matthew W. Wilson, and Matthew Zook. "Beyond the Geotag: Situating 'Big Data' and Leveraging the Potential of the Geoweb." *Cartography*

and Geographic Information Science 40, no. 2 (2013): 130–39. https://doi
.org/10.1080/15230406.2013.777137.

Crawford, Kate. *Atlas of AI: Power, Politics, and the Planetary Costs of Artificial Intelligence*. New Haven, CT: Yale University Press, 2021.

Cronon, William. *Nature's Metropolis: Chicago and the Great West*. 3rd ed. New York: W. W. Norton & Company, 1992.

Crunchbase. "List of Google's 259 Acquisitions, Including Photomath and Alter." Crunchbase. Accessed July 30, 2023. https://www.crunchbase.com/search /acquisitions/field/organizations/num_acquisitions/google.

Currid, Elizabeth. *The Warhol Economy*. Princeton, NJ: Princeton University Press, 2007.

Currid-Halkett, Elizabeth, and Allen J. Scott. "The Geography of Celebrity and Glamour: Reflections on Economy, Culture, and Desire in the City." *City, Culture and Society*, 2013.

Dalton, Craig M, Linnet Taylor, and Jim Thatcher. "Critical Data Studies: A Dialog on Data and Space." *Big Data & Society* 3, no. 1 (2016): 20539517 16648346. https://doi.org/10.1177/2053951716648346.

Dalton, Craig, and Jim Thatcher. "What Does a Critical Data Studies Look Like, and Why Do We Care? Seven Points for a Critical Approach to 'Big Data.'" *Society and Space*, May 12, 2014. https://www.societyandspace.org/articles /what-does-a-critical-data-studies-look-like-and-why-do-we-care.

Das, Tanmoy, CPJM van Elzakker, and Menno-Jan Kraak. "Conflicts in Neogeography Maps." In *Proceedings—AutoCarto 2012*, Columbus, OH, September 16–18, 2012.

Datta, Anusuya. "Landsat Completes 45 Years: Tracing the Journey." *Geospatial World* (blog), July 24, 2017. https://www.geospatialworld.net/blogs/landsat -completes-45-years/.

Davis, Mike. *City of Quartz: Excavating the Future in Los Angeles*. London: Verso, 2018.

———. *Ecology of Fear: Los Angeles and the Imagination of Disaster*. London: Verso, 2022.

Del Giudice, Vincent. "The government, saying it ran out of money, pulled. . . ." UPI Archives, March 3, 1989. https://www.upi.com/Archives/1989/03/02 /The-government-saying-it-ran-out-of-money-pulled/6994604818000/.

Dhaliwal, Ranjodh Singh. "On Addressability, or What Even Is Computation?" *Critical Inquiry* (2022). https://doi.org/10.1086/721167.

Diao, Mi, Hui Kong, and Jinhua Zhao. "Impacts of Transportation Network Companies on Urban Mobility." *Nature Sustainability* 4, no. 6 (June 2021): 494–500. https://doi.org/10.1038/s41893-020-00678-z.

Diario El Dia de La Plata. "Creen que la 'mafia de taxis' asesinó al marido de Georgina." April 11, 2001. https://www.eldia.com/nota/2001-11-4-creen-que -la-mafia-de-taxis-asesino-al-marido-de-georgina.

D'Ignazio, Catherine, and Lauren F. Klein. *Data Feminism*. Strong Ideas Series. Cambridge, MA: The MIT Press, 2020.

Dixon, A. D. "Variegated Capitalism and the Geography of Finance: Towards a Common Agenda." *Progress in Human Geography* 35, no. 2 (2011): 193–210. https://doi.org/10.1177/0309132510372006.

Dzieza, Josh. "Revolt of the Delivery Workers." *The Verge*, September 13, 2021. https://www.theverge.com/22667600/delivery-workers-seamless-uber-relay-new-york-electric-bikes-apps.

Economist. "The Cult of the Dabbawala." July 10, 2008. https://www.economist.com/business/2008/07/10/the-cult-of-the-dabbawala.

Elwood, Sarah. "Volunteered Geographic Information: Future Research Directions Motivated by Critical, Participatory, and Feminist GIS." *GeoJournal* 72 (2008): 173–83. https://doi.org/10.1007/s10708-008-9186-0.

Elwood, Sarah, and Agnieszka Leszczynski. "Privacy, Reconsidered: New Representations, Data Practices, and the Geoweb." *Geoforum* 42, no. 1 (January 2011): 6–15. https://doi.org/10.1016/j.geoforum.2010.08.003.

Elwood, Sarah, Nadine Schuurman, and Matthew W. Wilson. "Critical GIS." In *The SAGE Handbook of GIS and Society*, edited by Timothy L. Nyerges, Helen Couclelis, and Robert McMaster, 87–106. London: SAGE, 2011.

Emery, William, and Adriano Camps. "The History of Satellite Remote Sensing." In *Introduction to Satellite Remote Sensing*, 1–42. Amsterdam: Elsevier, 2017.

Erdmann, Eva. "Topographical Fiction: A World Map of International Crime Fiction." *Cartographic Journal* 48, no. 4 (November 1, 2011): 274–84. https://doi.org/10.1179/1743277411Y.0000000027.

European Space Agency. "Europe's Spaceport." Accessed July 20, 2023. https://www.esa.int/Enabling_Support/Space_Transportation/Europe_s_Spaceport/Europe_s_Spaceport2.

Federici, Silvia. *Caliban and the Witch*. Second, Revised edition. Brooklyn, NY: Autonomedia, 2014.

———. *Wages against Housework*. Bristol, UK: Falling Wall Press, 1975.

Fisher, Adam. "Google's Road Map to Global Domination." *New York Times Magazine*, December 11, 2013. http://www.nytimes.com/2013/12/15/magazine/googles-plan-for-global-domination-dont-ask-why-ask-where.html?_r=0.

Flores, Oscar. "Vandenberg AFB Renames Base and 30th Space Wing." NBC Los Angeles, *Space Force* (blog), May 12, 2021. https://www.nbclosangeles.com/news/local/vandenberg-afb-renames-base-and-30th-space-wing/2593887/.

Forelle, M. C. "The Material Consequences of 'Chipification': The Case of Software-Embedded Cars." *Big Data and Society* 9, no. 1 (January 1, 2022): 20539517221095429. https://doi.org/10.1177/20539517221095429.

Francica, Joe. "MapQuest Technology and Applications." *Directions Magazine*, March 26, 2004, 6.

Fraser, Nancy. *Fortunes of Feminism: From State-Managed Capitalism to Neoliberal Crisis*. Radical Thinkers. Brooklyn, NY: Verso Books, 2020.

Freidberg, Susanne Elizabeth. *Fresh: A Perishable History*. Cambridge, MA: Harvard University Press, 2009.

French, Robert L. "From Chinese Chariots to Smart Cars: 2,000 Years of Vehicular Navigation." *Navigation* 42, no. 1 (1995): 235–58. https://doi.org/10.1002/j.2161-4296.1995.tb02336.x.

Friedman, Thomas L. *The World Is Flat: A Brief History of the Twenty-First Century*. New York: Farrar, Straus and Giroux, 2005.

Gabrys, Jennifer. *Program Earth*. Minneapolis, MN: University of Minnesota Press, 2016. https://www.upress.umn.edu/book-division/books/program-earth.

Garfinkel, Perry. "Delivering Lunch in Mumbai, Across Generations." *New York Times*, February 2, 2017, sec. Job Market. https://www.nytimes.com/2017/02/02/jobs/dabbawalas-india-lunch.html.

Geoghegan, Bernard Dionysius. "The Bitmap Is the Territory: How Digital Formats Render Global Positions." *MLN* 136, no. 5 (2021): 1093–113. https://doi.org/10.1353/mln.2021.0081.

Gibson, C., and L. Kong. "Cultural Economy: A Critical Review." *Progress in Human Geography* 29, no. 5 (2005): 541–61.

Gibson, William. *Neuromancer*. Penguin, 2000.

Gibson-Graham, J. K. *The End of Capitalism (as We Knew It): A Feminist Critique of Political Economy*. Minneapolis: University of Minnesota Press, 2006.

Gieseking, Jen Jack. *A Queer New York: Geographies of Lesbians, Dykes, and Queers*. New York: New York University Press, 2020.

———. "An Everyday Queer New York." *Jgieseking.org* (blog), 2019. http://jgieseking.org/AQNY/AEQNYpubsmap/index.html.

Goetz, Marcus, and Alexander Zipf. "The Evolution of Geo-Crowdsourcing: Bringing Volunteered Geographic Information to the Third Dimension." In *Crowdsourcing Geographic Knowledge*, edited by Daniel Sui, Sarah Elwood, and Michael F. Goodchild, 139–58. Dordrecht: Springer, 2013.

Goldsmith, Jack, and Tim Wu. *Who Controls the Internet? Illusions of a Borderless World*. Oxford: Oxford University Press, 2006.

Goodchild, Michael F. "Citizens as Sensors: Web 2.0 and the Volunteering of Geographic Information." *GeoFocus*, no. 69 (2007): 211–21.

———. "Geographical Information Science." *International Journal of Geographical Information Systems* 6, no. 1 (1992): 31–45. https://doi.org/10.1080/02693799208901893.

Goodchild, Michael, Richard Appelbaum, Jeremy Crampton, William Herbert, Krzysztof Janowicz, Mei-Po Kwan, Katina Michael, et al. "A White Paper on Locational Information and the Public Interest." American Association of Geographers, September 2022. https://doi.org/10.14433/2017.0113.

Google. "Our Approach—How Google Search Works." Accessed June 17, 2022. https://www.google.com/search/howsearchworks/our-approach/.

Goranson, Christopher, Sayone Thihalolipavan, and Nicolás di Tada. "VGI and Public Health: Possibilities and Pitfalls." In *Crowdsourcing Geographic Knowledge: Volunteered Geographic Information (VGI) in Theory and Practice*, edited by Daniel Sui, Sarah Elwood, and Michael Goodchild, 329–40. Dordrecht: Springer, 2013.

Grabher, G., and O. Ibert. "Distance as Asset? Knowledge Collaboration in Hybrid Virtual Communities." *Journal of Economic Geography* 14, no. 1 (January 2014): 97–123. https://doi.org/10.1093/jeg/lbto14.

Graham, M., and T. Shelton. "Geography and the Future of Big Data, Big Data and the Future of Geography." *Dialogues in Human Geography* 3, no. 3 (2013): 255–61. https://doi.org/10.1177/2043820613513121.

Graham, Mark. "Time Machines and Virtual Portals: The Spatialities of the Digital Divide." *Progress in Development Studies* 11, no. 3 (2011): 211–27.

Graham, Mark, and Martin Dittus. *Geographies of Digital Exclusion: Data and Inequality*. London: Pluto Press, 2022.

Graham, Stephen. *Cities under Siege: The New Military Urbanism*. London: Verso, 2011.

———. "Software-Sorted Geographies." *Progress in Human Geography* 29, no. 5 (2005): 562–80. https://doi.org/10.1191/0309132505ph568oa.

———. "The End of Geography or the Explosion of Place? Conceptualizing Space, Place and Information Technology." *Progress in Human Geography* 22, no. 2 (April 1, 1998): 165–85. https://doi.org/10.1191/030913298673134137.

Graham, Stephen, and Simon Marvin. *Splintering Urbanism: Networked Infrastructures, Technological Mobilities and the Urban Condition*. London: Routledge, 2001.

———. *Telecommunications and the City: Electronic Spaces, Urban Places*. London: Routledge, 2002. http://ebookcentral.proquest.com/lib/qut/detail.action?docID=166821.

Granovetter, Mark. *Society and Economy*. Cambridge, MA: Harvard University Press, 2017.

Greene, Daniel. "Landlords of the Internet: Big Data and Big Real Estate." *Social Studies of Science* 52, no. 6 (2022): 904–27. https://doi.org/10.1177/03063127221124943.

———. *The Promise of Access: Technology, Inequality, and the Political Economy of Hope*. Cambridge, MA: MIT Press, 2021.

Grubhub. "Getting Paid as a Grubhub Driver." Grubhub for Drivers, 2021. https://driver.grubhub.com/pay/.

Guardian. "The Uber Files." July 2022. https://www.theguardian.com/news/series/uber-files.

Habermann, Ina, and Nikolaus Kuhn. "'Sustainable Fictions' Geographical, Literary and Cultural Intersections in J. R. R. Tolkien's *The Lord of the Rings*." *Cartographic Journal* 48, no. 4 (2011): 263–73. https://doi.org/10.1179/1743277411Y.0000000024.

Haklay, Mordechai Muki. "Neogeography and the Delusion of Democratisation." *Environment and Planning A* 45, no. 1 (2013): 55–69. https://doi.org/10.1068/a45184.

———. "Citizen Science and Volunteered Geographic Information: Overview and Typology of Participation." In *Crowdsourcing Geographic Knowledge*, edited by Daniel Sui, Sarah Elwood, and Michael F. Goodchild, 105–20. Dordrecht: Springer, 2013.

Hall, Peter. "Creative Cities and Economic Development." *Urban Studies* 37 (1999): 639–49.

Hambleton, Kathryn. "Around the Moon with NASA's First Launch of SLS with Orion." NASA, March 7, 2018. http://www.nasa.gov/feature/around-the-moon-with-nasa-s-first-launch-of-sls-with-orion.

———. "Artemis I Overview." NASA, February 20, 2018. http://www.nasa.gov/content/artemis-i-overview.

Harlan, Chico. "'Does MapQuest Still Exist?' Yes, It Does, and It's a Profitable Business." *Washington Post*, May 22, 2015, sec. Business. https://www.washingtonpost.com/business/economy/does-mapquest-still-exist-as-a-matter-of-fact-it-does/2015/05/22/995d2532-fa5d-11e4-a13c-193b1241d51a_story.html.

Harley, J. B. "Deconstructing the Map." *Cartographica: The International Journal for Geographic Information and Geovisualization* 26, no. 2 (1989): 1–20. https://doi.org/10.3138/E635-7827-1757-9T53.

Harvey, David. "Reflections on an Academic Life." *Human Geography* 15, no. 1 (2022): 14–24. https://doi.org/10.1177/19427786211046291.

———. *The Condition of Postmodernity: An Enquiry into the Origins of Cultural Change*. Malden, MA: Blackwell, 1989.

Hicks, Marie. *Programmed Inequality*. Cambridge, MA: MIT Press, 2017. https://mitpress.mit.edu/books/programmed-inequality.

Hollingham, Richard. "V2: The Nazi Rocket That Launched the Space Age." BBC, September 7, 2014. https://www.bbc.com/future/article/20140905-the-nazis-space-age-rocket.

Houri, Cyril. Method and systems for locating geographical locations of online users. United States US6665715B1, filed April 3, 2000, and issued December 16, 2003. https://patents.google.com/patent/US6665715B1/en?assignee=cyril+houri&oq=cyril+houri&sort=old.

Hu, Tung-Hui. *A Prehistory of the Cloud*. Cambridge, MA: MIT Press, 2015.

Hussain, Suhauna. "Prop. 22 Was Ruled Unconstitutional. What Will the Final Outcome Be?" *Los Angeles Times*, August 25, 2021. https://www.latimes.com/business/technology/story/2021-08-25/after-prop-22-ruling-whats-next-uber-lyft.

Ibeh, Joseph. "Kenya and Italy Close In on Signing Ownership Deal in Respect of Luigi Broglio Space Centre." *Space in Africa* (blog), July 19, 2019. https://africanews.space/kenya-and-italy-signing-ownership-deal-luigi-broglio-space-centre/.

ICEYE. "A Revolution in Synthetic Aperture Radar (SAR) Data Earth Observation." 2022. https://www.iceye.com/hubfs/Downloadables/SAR_Data_Brochure_ICEYE.pdf.

Indap, Sujeet. "Wall Street Eyes NYC Taxis as Beleaguered Drivers Win Relief." *Financial Times*, November 15, 2021.

Indian Space Research Organization, and Antrix Corporation Limited. "PSLV-C37 Brochure." ISRO, 2017. https://www.isro.gov.in/pslv-c37-cartosat-2-series-satellite/pslv-c37-brochure-0.

International Civil Aviation Organization. "The World of Air Transport in 2019." 2019. https://www.icao.int/annual-report-2019/Pages/the-world-of-air-transport-in-2019.aspx.

———. "Economic Impacts of COVID-19 on Civil Aviation." ICAO, June 10, 2022. https://www.icao.int/sustainability/Pages/Economic-Impacts-of-COVID-19.aspx.

International Energy Agency, and International Union of Railways. *The Future of Rail: Opportunities for Energy and the Environment*. IEA, 2019. https://doi.org/10.1787/9789264312821-en.

Iveson, Kurt, and Sophia Maalsen. "Social Control in the Networked City: Data-fied Dividuals, Disciplined Individuals and Powers of Assembly." *Environment and Planning D: Society and Space* 37, no. 2 (2019): 331–49. https://doi .org/10.1177/0263775818812084.

Izushi, Hiro, and Yuko Aoyama. "Industry Evolution and Cross-Sectoral Skill Transfers: A Comparative Analysis of the Video Game Industry in Japan, the United States, and the United Kingdom." *Environment and Planning A: Economy and Space* 38 (2006): 1843–61. https://doi.org/10.1068/a37205.

Jacob, John P., Luke Skrebowski, and Smithsonian American Art Museum, eds. *Trevor Paglen: Sites Unseen.* Washington, DC; London: Smithsonian American Art Museum in association with D Giles Limited, 2018.

Jacobsen, Annie. *Operation Paperclip: The Secret Intelligence Program That Brought Nazi Scientists to America.* New York: Little Brown and Company, 2015.

Jarrett, Kylie. "The Relevance of 'Women's Work': Social Reproduction and Immaterial Labor in Digital Media." *Television & New Media* 15, no. 1 (2014): 14–29. https://doi.org/10.1177/1527476413487607.

Jefferson, Brian Jordan. *Digitize and Punish: Racial Criminalization in the Digital Age.* Minneapolis: University of Minnesota Press, 2020.

Jessop, Bob. "Rethinking the Diversity of Capitalism: Varieties of Capitalism, Variegated Capitalism, and the World Market." In *Capitalist Diversity and Diversity within Capitalism*, edited by Geoffrey Wood and Christel Lane, 209–37. London: Routledge, 2011. http://eprints.lancs.ac.uk/55149/.

Johnson, Peter A., and Renee E. Sieber. "Situating the Adoption of VGI by Government," edited by Daniel Sui, Sarah Elwood, and Michael F. Goodchild, 65–81. Dordrecht: Springer, 2013.

Kassens-Noor, E., Josh Siegel, and Travis Decaminada. "Choosing Ethics Over Morals: A Possible Determinant to Embracing Artificial Intelligence in Future Urban Mobility." *Frontiers in Sustainable Cities* 28, no. 3 (2021). https://www .frontiersin.org/articles/10.3389/frsc.2021.723475.

Kitchin, Rob. *The Data Revolution.* Los Angeles: SAGE, 2014.

Kitchin, Rob, and Martin Dodge. *Code/Space: Software and Everyday Life.* Cambridge, MA.: MIT Press, 2011.

Kitchin, Rob, and Tracey P. Lauriault. "Towards Critical Data Studies: Charting and Unpacking Data Assemblages and Their Work." *The Programmable City Working Paper*, 2014. http://papers.ssrn.com/sol3/papers.cfm?abstract_id= 2474112.

Kogler, Dieter F., David L. Rigby, and Isaac Tucker. "Mapping Knowledge Space and Technological Relatedness in US Cities." *European Planning Studies* 21, no. 9 (2013): 1374–91. https://doi.org/10.1080/09654313.2012.755832.

Kooser, Amanda. "Elon Musk Breaks Down the Starship Numbers for a Million-Person SpaceX Mars Colony." CNET, January 16, 2020. https://www.cnet .com/science/elon-musk-drops-details-for-spacexs-million-person-mars-mega -colony/.

Kwan, Mei-Po. "Introduction: Feminist Geography and GIS." *Gender, Place & Culture* 9, no. 3 (2002): 261–62. https://doi.org/10.1080/096636902200 0003860.

———. "Feminist Visualization: Re-Envisioning GIS as a Method in Feminist Geographic Research." *Annals of the Association of American Geographers* 92, no. 4 (2002): 645–61.

La Nación. "Asesinan en un taxi en Palermo al marido de Georgina Barbarossa." November 3, 2001. https://www.lanacion.com.ar/sociedad/asesinan-en-un-taxi-en-palermo-al-marido-de-georgina-barbarossa-nid348353/.

———. "El taxista que atropelló a Leonela no dijo que era diabético cuando sacó el registro." February 26, 2013. https://www.lanacion.com.ar/seguridad/el-taxista-que-atropello-a-leonela-no-dijo-que-era-diabetico-cuando-saco-el-registro-nid1558234/.

Lally, Nick, Kelly Kay, and Jim Thatcher. "Computational Parasites and Hydropower: A Political Ecology of Bitcoin Mining on the Columbia River." *Environment and Planning E: Nature and Space* 5, no. 1 (2019). https://doi.org/10.1177/2514848619867608.

Langley, Paul, and Andrew Leyshon. "Platform Capitalism: The Intermediation and Capitalization of Digital Economic Circulation." *Finance and Society* 3, no. 1 (2017): 11–31. https://doi.org/10.2218/finsoc.v3i1.1936.

Lasswell, Harold Dwight. *Politics: Who Gets What, When, How.* Whitefish, MT: Literary Licensing, 2013.

Latam Satelital. "Venesat-1 queda fuera de servicio." March 29, 2020. http://latamsatelital.com/venesat-1-queda-fuera-de-servicio/.

Leamer, Edward E., and Michael Storper. "The Economic Geography of the Internet Age." *NBER Working Paper Series*, no. 8450 (August 2001).

Légifrance. "Code Pénal, Article R645-1." Accessed August 7, 2022. https://www.legifrance.gouv.fr/codes/article_lc/LEGIARTI000022375941.

Leslie, Deborah, and Norma M. Rantisi. "Creativity and Place in the Evolution of a Cultural Industry: The Case of Cirque Du Soleil." *Urban Studies* 48, no. 1 (July 2011): 1771–87.

Leszczynski, Agnieszka. "[Digital] Spatialities." In *Digital Geographies*, edited by James Ash, Rob Kitchin, and Agnieszka Leszczynski, 13–23. Thousand Oaks, CA: SAGE Publications, 2018.

———. "Situating the Geoweb in Political Economy." *Progress in Human Geography* 36, no. 1 (2012): 72–89. https://doi.org/10.1177/0309132511411231.

———. "Spatial Big Data and Anxieties of Control." *Environment and Planning D: Society and Space* 33, no. 6 (2015): 965–84. https://doi.org/10.1177/0263775815595814.

Leszczynski, Agnieszka, and Matthew W. Wilson. "Guest Editorial: Theorizing the Geoweb." *GeoJournal* 78 (July 12, 2013): 915–19. https://doi.org/10.1007/s10708-013-9489-7.

Lewis, Elspeth. "How Sputnik Changed the World." National Space Centre, March 10, 2017. https://spacecentre.co.uk/blog-post/sputnik-changed-world/.

Lin, Cindy. "How Forest Became Data: The Remaking of Ground Truth in Indonesia." In *The Nature of Data: Infrastructures, Environments, Politics*, edited by Jenny Goldstein and Eric Nost, 285–302. Lincoln: University of Nebraska Press, 2022.

Lippitt, Christopher. "Georeferencing and Georectification." *Geographic Information Science & Technology Body of Knowledge*, no. Q3 (2020). https://doi.org/10.22224/gistbok/2020.3.3.

Lipson, Hod, and Melba Kurman. *Driverless: Intelligent Cars and the Road Ahead*. Cambridge, MA: MIT Press, 2016.

Lorenzen, M., and R. Mudambi. "Clusters, Connectivity and Catch-up: Bollywood and Bangalore in the Global Economy." *Journal of Economic Geography* 13, no. 3 (May 2013): 501–34. https://doi.org/10.1093/jeg/lbs017.

Lutkevich, Ben, and John Burke. "What Is DNS? How Domain Name System Works." TechTarget Network Infrastructure, August 2021. https://www.techtarget.com/searchnetworking/definition/domain-name-system.

M. "The Space Center Kenya Doesn't Own." Owaahh, March 31, 2016. https://owaahh.com/space-center-kenya-doesnt/.

Mabrouk, Elizabeth. "What Are SmallSats and CubeSats?" NASA, March 13, 2015. http://www.nasa.gov/content/what-are-smallsats-and-cubesats.

Macartney, Huw. *Variegated Neoliberalism: EU Varieties of Capitalism and International Political Economy*. London: Taylor & Francis, 2010.

Mack, Pamela Etter. *Viewing the Earth: The Social Construction of the Landsat Satellite System*. Inside Technology. Cambridge, MA: MIT Press, 1990.

Mailland, Julien, and Kevin Driscoll. *Minitel: Welcome to the Internet*. Cambridge, MA: MIT Press, 2017.

Malecki, Edward J. "Digital Development in Rural Areas: Potentials and Pitfalls." *Journal of Rural Studies* 19 (2003): 201–14. https://doi.org/10.1016/s0743-0167(02)00068-2.

Malecki, Edward J., and Bruno Moriset. "Organization versus Space: The Paradoxical Geographies of the Digital Economy." *Geography Compass* 3, no. 1 (2009): 256–74. https://doi.org/10.1111/j.1749-8198.2008.00203.x.

———. *The Digital Economy*. Abingdon, UK: Routledge, 2007.

Marx, Paris. *Road to Nowhere: What Silicon Valley Gets Wrong about the Future of Transportation*. London: Verso, 2022.

Massey, Doreen. "A Global Sense of Place." In *Space, Place, and Gender*, 146–56. Minneapolis: University of Minnesota Press, 1994. http://ebookcentral.proquest.com/lib/dartmouth-trial/detail.action?docID=310284.

May, Martha. "The Historical Problem of the Family Wage: The Ford Motor Company and the Five Dollar Day." *Feminist Studies* 8, no. 2 (1982): 399–424. https://doi.org/10.2307/3177569.

Mayer III, Stephen. *The Five Dollar Day: Labor Management and Social Control in the Ford Motor Company, 1908–1921*. Albany: State University of New York Press, 1981.

Medina, Eden. *Cybernetic Revolutionaries: Technology and Politics in Allende's Chile*. Cambridge, MA: MIT Press, 2014. https://doi.org/10.7551/mitpress/8417.001.0001.

———. "Designing Freedom, Regulating a Nation: Socialist Cybernetics in Allende's Chile." *Journal of Latin American Studies* 38, no. 3 (2006): 571–606. https://doi:10.1017/S0022216X06001179.

Mosco, Vincent. *The Digital Sublime: Myth, Power, and Cyberspace*. Cambridge, MA: MIT Press, 2005.

Mulvaney, Erin, and Kathleen Dailey. "Will Uber's U.K. Loss Jump the Pond? Gig Worker Status Explained." *Bloomberg Law*, February 22, 2021. https://news.bloomberglaw.com/daily-labor-report/will-ubers-u-k-loss-jump-the-pond-gig-worker-status-explained.

NASA. "Agency Awards Historical Recipient List." https://searchpub.nssc.nasa.gov/servlet/sm.web.Fetch?rhid=1000&did=2120817&type=released.

———. "Artemis." Accessed August 29, 2022. https://www.nasa.gov/specials/artemis/index.html.

———. "SP-4012 NASA Historical Data Book: Volume IV. NASA Resources 1969-1978." Accessed August 28, 2022. https://history.nasa.gov/SP-4012/vol4/appa.htm.

———. "Value of NASA." NASA. Accessed July 20, 2023. https://www.nasa.gov/specials/value-of-nasa/index.html.

NASA Jet Propulsion Laboratory (JPL), California Institute of Technology. "Seasat—Earth Missions." Accessed August 25, 2022. https://www.jpl.nasa.gov/missions/seasat.

———. "Trailblazer Sea Satellite Marks Its Coral Anniversary." June 27, 2013. https://www.jpl.nasa.gov/news/trailblazer-sea-satellite-marks-its-coral-anniversary.

National Archives. "Records of the Secretary of Defense (RG 330)." August 15, 2016. https://www.archives.gov/iwg/declassified-records/rg-330-defense-secretary.

National Geospatial Advisory Committee Landsat Advisory Group. "Revisiting the Land Remote Sensing Policy Act of 1992." NGAC, April 2021. https://www.fgdc.gov/ngac/meetings/april-2021/ngac-paper-revisiting-the-land-remote-sensing.pdf.

National Reconnaissance Office. "The CORONA Program." Accessed August 27, 2022. https://www.nro.gov/History-and-Studies/Center-for-the-Study-of-National-Reconnaissance/The-CORONA-Program/.

Neufeld, Michael J. *The Rocket and the Reich: Peenemunde and the Coming of the Ballistic Missile Era*. Washington D.C.: Smithsonian Institution, 2013.

Nido, Juan Manuel del. *Taxis vs. Uber: Courts, Markets, and Technology in Buenos Aires*. Stanford, CA: Stanford University Press, 2021. https://doi.org/10.1515/9781503629684.

Noble, Safiya Umoja. *Algorithms of Oppression: How Search Engines Reinforce Racism*. New York: New York University Press, 2018.

Norcliffe, Glen, and Olivero Rendace. "New Geographies of Comic Book Production in North America: The New Artisan, Distancing, and the Periodic Social Economy." *Economic Geography* 79, no. 3 (2003): 241–63. https://doi.org/10.1111/j.1944-8287.2003.tb00211.x.

On-Road Automated Driving (ORAD) Committee. "Taxonomy and Definitions for Terms Related to Driving Automation Systems for On-Road Motor Vehicles." SAE International, April 30, 2021. https://doi.org/10.4271/J3016_202104.

Openshaw, Stan. "A View on the GIS Crisis in Geography; or, Using GIS to Put Humpty-Dumpty Back Together Again." *Environment and Planning A* 23, no. 5 (1991): 621–28.

———. "The Truth about Ground Truth." *Transactions in GIS* 2, no. 1 (1997): 7–24.

O'Reilly, Tim. "What Is Web 2.0: Design Patterns and Business Models for the Next Generation of Software." *Communications and Strategies* 65, no. 1 (January 1, 2007): 17–37.

O'Sullivan, David, and George L. W. Perry. "Pattern, Process and Scale." In *Spatial Simulation: Exploring Pattern and Process*, 29–56. Chichester, UK: John Wiley & Sons, Ltd, 2013.

Ozersky, Josh. *The Hamburger: A History*. New Haven, CT: Yale University Press, 2008.

Palmer, Mark H., and Scott K. Kraushaar. "Volunteered Geographic Information, Actor-Network Theory, and Severe-Storm Reports," edited by Daniel Sui, Sarah Elwood, and Michael F. Goodchild, 287–306. Dordrecht: Springer, 2013.

Pathak, Gauri Sanjeev. "Delivering the Nation: The Dabbawalas of Mumbai." *South Asia: Journal of South Asian Studies* 33, no. 2 (2010): 235–57. https://doi.org/10.1080/00856401.2010.493280.

Peck, Jamie, and Nik Theodore. "Variegated Capitalism." *Progress in Human Geography* 31, no. 6 (2007): 731–72. https://doi.org/10.1177/030913250507 83505.

Peck, Jamie, Nik Theodore, and Neil Brenner. "Neoliberal Urbanism: Models, Moments, Mutations." *SAIS Review* 29, no. 1 (2009): 49–66.

Perkins, Richard, and Eric Neumayer. "Is the Internet Really New After All? The Determinants of Telecommunications Diffusion in Historical Perspective." *Professional Geographer* 63 (2011): 55–72.

Petchenik, Barbara B. "Donnelley Cartographic Services." *American Cartographer* 14, no. 3 (January 1987): 241–44. https://doi.org/10.1559/15230408 7783875796.

Peters, Benjamin. *How Not to Network a Nation: The Uneasy History of the Soviet Internet*. Cambridge, MA: MIT Press, 2017.

Petersen, Michael B. *Missiles for the Fatherland: Peenemünde, National Socialism, and the V-2 Missile*. Cambridge Centennial of Flight. Cambridge: Cambridge University Press, 2011.

Peterson, Michael P. "MapQuest and the Beginnings of Web Cartography." *International Journal of Cartography* 7, no. 2 (2021): 275–81. https://doi.org/10 .1080/23729333.2021.1925831.

Petralia, Sergio, Pierre-Alexandre Balland, and David L. Rigby. "Unveiling the Geography of Historical Patents in the United States from 1836 to 1975." *Scientific Data* 3 (2016): 160074. https://doi.org/10.1038/sdata.2016.74.

Pickles, John. "Arguments, Debates, and Dialogues: The GIS-Social Theory Debate and the Concern for Alternatives." *Geographical Information Systems* 1 (1999): 49–60.

———. *Ground Truth: The Social Implications of Geographic Information Systems*. New York: Guilford Press, 1995.

Pisacane, Vincent L. "The Legacy of Transit: A Dedication." *Johns Hopkins APL Technical Digest* 19, no. 1 (1998): 5–10.

Planet Labs PBC. "Company." Planet, 2022. https://www.planet.com/company/.

Porter, M. "Location, Competition, and Economic Development: Local Clusters in a Global Economy." *Economic Development Quarterly* 14 (2000): 15–34.

Powell, Walter W., and Kaisa Snellman. "The Knowledge Economy." *Annual Review of Sociology* 30 (2004): 199–220.

Power, Dominic, and Allen J. Scott. *Cultural Industries and the Production of Culture.* Psychology Press, 2004.

Queering The Map. "Queering The Map." Accessed July 14, 2022. https://www.queeringthemap.com/.

Quinn, Sterling, and Luis F. Alvarez León. "Every Single Street? Rethinking Full Coverage across Street-Level Imagery Platforms." *Transactions in GIS*, 23, no. 6 (2019): 1251–72. https://doi.org/10.1111/tgis.12571.

Raff, Daniel M. G., and Lawrence H. Summers. "Did Henry Ford Pay Efficiency Wages?" *Journal of Labor Economics* 5, no. 4 (October 2, 1987): s57–86.

Rankin, William. *After the Map: Cartography, Navigation, and the Transformation of Territory in the Twentieth Century.* Chicago: University of Chicago Press, 2018. https://press.uchicago.edu/ucp/books/book/chicago/A/bo22655244.html.

Redfield, Peter. *Space in the Tropics: From Convicts to Rockets in French Guiana.* Berkeley: University of California Press, 2000.

Rettig, Molly. "Old Satellite Imagery Offers New Baseline Data." *Anchorage Daily News*, June 22, 2013. https://archive.ph/7PqBy.

Richterich, Annika. "Cartographies of Digital Fiction: Amateurs Mapping a New Literary Realism." *Cartographic Journal* 48, no. 4 (November 1, 2011): 237–49. https://doi.org/10.1179/1743277411Y.0000000021.

Richwine, Lisa, and Dawn Chmielewski. "Hollywood Writers Strike over Pay in Streaming TV 'Gig Economy.'" Reuters, May 2, 2023. https://www.reuters.com/lifestyle/hollywood-writers-studios-stage-last-minute-talks-strike-deadline-looms-2023-05-01/.

Ritchie, Hannah. "Cars, Planes, Trains: Where Do CO2 Emissions from Transport Come From?" Our World in Data, October 6, 2020. https://ourworldindata.org/co2-emissions-from-transport.

Rosen, Jovanna, and Luis F. Alvarez León. "Signaling Hinterlands and the Spatial Networks of Digital Capitalism." *Annals of the American Association of Geographers* (2023): 1–13. https://doi.org/10.1080/24694452.2023.2249974.

———. "The Digital Growth Machine: Urban Change and the Ideology of Technology." *Annals of the American Association of Geographers* 112, no. 8, (2022): 2248–2265. https://doi.org/10.1080/24694452.2022.2052008.

Rosen, Julia. "Shifting Ground: Fleets of Radar Satellites Are Measuring Movements on Earth Like Never Before." *Science*, February 25, 2021. https://doi.org/10.1126/science.abh2435.

Rosenblat, Alex. *Uberland: How Algorithms Are Rewriting the Rules of Work.* Oakland: University of California Press, 2018.

Rosenthal, Brian M. "N.Y.C. Cabbies Win Millions More in Aid After Hunger Strike." *New York Times*, November 3, 2021, sec. New York. https://www.nytimes.com/2021/11/03/nyregion/nyc-taxi-drivers-hunger-strike.html.

Rottenberg, Josh. "Hollywood Actors Join WGA in Historic Double Strike. 'This Is All of Our Fight.'" *Los Angeles Times*, July 14, 2023. https://www.latimes.com/entertainment-arts/business/story/2023-07-14/actors-strike-sag-aftra-joins-writers-guild-picket-lines.

Sadowski, Jathan. *Too Smart: How Digital Capitalism Is Extracting Data, Controlling Our Lives, and Taking Over the World*. Cambridge, MA: MIT Press, 2020.

Salam, Erum. "'They Stole from Us': The New York Taxi Drivers Mired in Debt over Medallions." *Guardian*, October 2, 2021, sec. US news. https://www.theguardian.com/us-news/2021/oct/02/new-york-city-taxi-medallion-drivers-debt.

Satellite Imaging Corporation. "Planetscope—Dove Satellite Constellation (3m)." 2022. https://www.satimagingcorp.com/satellite-sensors/other-satellite-sensors/dove-3m/.

Saxon, Wolfgang. "Arthur Rudolph, 89, Developer of Rocket in First Apollo Flight." *New York Times*, January 3, 1996, sec. U.S. https://www.nytimes.com/1996/01/03/us/arthur-rudolph-89-developer-of-rocket-in-first-apollo-flight.html.

Scharl, Arno, and Klaus Tochtermann. *The Geospatial Web*. London: Springer, 2007.

Schoon, Ben. "Numbers from Google I/O: 3x Growth for Wear OS, 3 Billion Active Android Devices, More." *9to5Google* (blog), May 12, 2022. https://9to5google.com/2022/05/11/google-io-2022-numbers/.

Schuurman, Nadine. "Trouble in the Heartland: GIS and Its Critics in the 1990s." *Progress in Human Geography* 24, no. 4 (2000): 569–90.

Schuurman, Nadine, and Geraldine Pratt. "Care of the Subject: Feminism and Critiques of GIS." *Gender, Place & Culture* 9, no. 3 (2002): 291–99. https://doi.org/10.1080/0966369022000003905.

Scott, Allen J. "Capitalism and Urbanization in a New Key? The Cognitive-Cultural Dimension." *Social Forces* 85, no. 4 (2007): 1465–82.

———. *The Cultural Economy of Cities: Essays on the Geography of Image-Producing Industries*. London: SAGE, 2000.

———. "Economic Geography: The Great Half-Century." *Cambridge Journal of Economics* 24, no. 4 (2000): 483–504.

———. *On Hollywood: The Place, The Industry*. Princeton, NJ: Princeton University Press, 2005.

———. *Social Economy of the Metropolis: Cognitive-Cultural Capitalism and the Global Resurgence of Cities*. Oxford University Press, 2008.

Sellier, Andre. *A History of the Dora Camp: The Untold Story of the Nazi Slave Labor Camp That Secretly Manufactured V-2 Rockets*. Chicago: Ivan R. Dee, 2003.

Sheller, Mimi. *Mobility Justice: The Politics of Movement in the Age of Extremes*. London: Verso, 2018.

Sheller, Mimi, and John Urry. "The City and the Car." *International Journal of Urban and Regional Research* 24, no. 4 (December 2000): 737–57. https://doi.org/10.1111/1468-2427.00276.

Sheppard, Eric S. "Knowledge Production through Critical GIS: Genealogy and Prospects." *Cartographica: The International Journal for Geographic Information* 40, no. 4 (Winter 2005): 5–21.

———. *Limits to Globalization: Disruptive Geographies of Capitalist Development*. Oxford: Oxford University Press, 2016.

Skelton, Sebastian Klovig. "Uber Drivers Strike over Pay Issues and Algorithmic Transparency." *ComputerWeekly*, July 22, 2022. https://www.computerweekly.com/news/252521871/Uber-drivers-strike-over-pay-issues-and-algorithmic-transparency.

Smith, Harrison. "Open and Free? The Political Economy of the Geospatial Web 2.0." *Geothink Working Paper Series*, Geothink Working Paper Series, no. 001 (2014). http://geothink.ca/wp-content/uploads/2014/06/Geothink-Working-Paper-001-Shade-Smith1.pdf.

Smith, Neil. "History and Philosophy of Geography: Real Wars, Theory Wars." *Progress in Human Geography* 16, no. 2 (1992): 257–71.

SpaceTech Analytics. "SpaceTech Industry 2021 / Q2 Landscape Overview." May 2021.

Srnicek, Nick. *Platform Capitalism*. Cambridge, UK: Polity Press, 2017.

Starosielski, Nicole. *The Undersea Network*. Sign, Storage, Transmission. Durham, NC: Duke University Press, 2015.

Statista. "Seaborne Trade—Capacity of Container Ships 2021." November 2021. https://www.statista.com/statistics/267603/capacity-of-container-ships-in-the-global-seaborne-trade/.

Steinberg, Marc. "From Automobile Capitalism to Platform Capitalism: Toyotism as a Prehistory of Digital Platforms." *Organization Studies* 43, no. 7 (2022): 1069–90. https://doi.org/10.1177/0170840621103068.

Stokes, Mark, Gabriel Alvarado, Emily Weinstein, and Ian Easton. "China's Space and Counterspace Capabilities and Activities." US-China Economic and Security Review Commission, March 30, 2020. https://www.uscc.gov/research/chinas-space-and-counterspace-activities.

Storper, Michael, and Susan Christopherson. "Flexible Specialization and Regional Industrial Agglomerations: The Case of the U.S. Motion Picture Industry." *Annals of the Association of American Geographers* 77, no. 1 (1987): 104–17.

Straits Research. "Satellite Data Services Market Size, Share and Analysis 2031." Satellite Data Services Market, 2023. https://straitsresearch.com/report/satellite-data-services-market.

Sui, Daniel, Sarah Elwood, and Michael Goodchild. *Crowdsourcing Geographic Knowledge*. Dordrecht: Springer, 2013.

Superior Court of Paris. *LICRA and UEJF vs YAHOO! Inc. and YAHOO FRANCE*. May 22, 2000. http://www.lapres.net/yahen.html.

Tarnoff, Ben. *Internet for the People: The Fight for Our Digital Future*. London: Verso, 2022.

Tufekci, Zeynep. *Twitter and Tear Gas: The Power and Fragility of Networked Protest*. New Haven, CT: Yale University Press, 2017.

Tung, Stephen. "How the Finnish School System Outshines U.S. Education." Stanford Report, January 20, 2012. http://news.stanford.edu/news/2012/january/finnish-schools-reform-012012.html.

Turner, Andrew. *Introduction to Neogeography*. Sebastopol, CA: O'Reilly Media, Inc., 2006.

Turner, Fred. *From Counterculture to Cyberculture: Stewart Brand, the Whole Earth Network, and the Rise of Digital Utopianism*. Chicago: University of Chicago Press, 2008.

Turner, Matthew D., and Peter J. Taylor. "Critical Reflections on the Use of Remote Sensing and GIS Technologies in Human Ecological Research." *Human Ecology* 31, no. 2 (June 2003): 177–82. https://doi.org/10.1023/A:1023958712140.

Uber. "Uber Cities—Rides Around the World." 2022. https://www.uber.com/global/en/cities/.

United States Congress. Land Remote Sensing Policy Act of 1992, Pub. L. No. 102-555 (1992). https://www.govinfo.gov/app/details/https%3A%2F%2Fwww.govinfo.gov%2Fapp%2Fdetails%2FCOMPS-1849.

University College Cork. "Irish Emigration History." Accessed May 22, 2023. https://www.ucc.ie/en/emigre/history/#_ftn8.

University of Chicago Library. "Guide to the R.R. Donnelley & Sons Company Archive 1844–2005." https://www.lib.uchicago.edu/e/scrc/findingaids/view.php?eadid=ICU.SPCL.DONNELLEY.

University of Wisconsin Department of Geography. "History of Cartography Project." Accessed July 22, 2022. https://geography.wisc.edu/histcart/.

Urry, John. "The 'System' of Automobility." *Theory, Culture & Society* 21, nos. 4–5 (2004): 25–39. https://doi.org/10.1177/0263276404046059.

US Geological Survey. "What Does 'Georeferenced' Mean?" Accessed July 25, 2022. https://www.usgs.gov/faqs/what-does-georeferenced-mean.

US Geological Survey's Earth Resources Observation and Science (EROS) Center. "Products Overview." *USGS EROS Archive* (blog), July 19, 2019. https://www.usgs.gov/centers/eros/science/usgs-eros-archive-products-overview.

Wade, Mark. "R-7." *Encyclopedia Astronautica*, September 4, 2003. https://web.archive.org/web/20030904120332/http://www.astronautix.com/lvs/r7.htm.

Warf, Barney, and Daniel Sui. "From GIS to Neogeography: Ontological Implications and Theories of Truth." *Annals of GIS* 16, no. 4 (2010): 197–209. https://doi.org/10.1080/19475683.2010.539985.

Waring, Marilyn. "Counting for Something! Recognising Women's Contribution to the Global Economy through Alternative Accounting Systems." *Gender & Development* 11, no. 1 (2003): 35–43. https://doi.org/10.1080/741954251.

Wells, Katie J., Kafui Ablode Attoh, and Declan Cullen. *Disrupting D.C.: The Rise of Uber and the Fall of the City*. Princeton, NJ: Princeton University Press, 2023.

Wells, Katie J., Kafui Attoh, and Declan Cullen. "'Just-in-Place' Labor: Driver Organizing in the Uber Workplace." *Environment and Planning A: Economy and Space* 53, no. 2 (2021): 315–31. https://doi.org/10.1177/0308518X20949266.

Wilford, John Noble. "US Halts Plan to Turn Off the Landsat Satellites." *New York Times*, March 17. 1989.

Wilson, Matthew W. *New Lines: Critical GIS and the Trouble of the Map*. Minneapolis: University of Minnesota Press, 2017.

———. "New Lines? Enacting a Social History of GIS." *The Canadian Geographer / Le Géographe Canadien* 59, no. 1 (2015): 29–34. https://doi.org/10.1111/cag.12118.

———. "Situating Neogeography." *Environment and Planning A* 45, no. 1 (2013): 3–9.

Wilson, Matthew W., and Mark Graham. "Neogeography and Volunteered Geographic Information: A Conversation with Michael Goodchild and Andrew Turner." *Environment and Planning A* 45, no. 1 (2013): 10–18. https://doi.org/10.1068/a44483.

Wired staff. "Clinton Unscrambles GPS Signals." *Wired*, May 1, 2000. https://www.wired.com/2000/05/clinton-unscrambles-gps-signals/.

World Bank. "Net Migration—Ireland." World Bank Open Data, 2022. https://data.worldbank.org.

Wu, Tim. *The Attention Merchants: The Epic Scramble to Get inside Our Heads*. New York: Vintage Books, 2017.

———. *The Master Switch: The Rise and Fall of Information Empires*. New York: Vintage Books, 2011.

Yeong, De Jong, Gustavo Velasco-Hernandez, John Barry, and Joseph Walsh. "Sensor and Sensor Fusion Technology in Autonomous Vehicles: A Review." *Sensors* 21, no. 6 (January 2021): 2140. https://doi.org/10.3390/s21062140.

Yoon, H., and E. J. Malecki. "Cartoon Planet: Worlds of Production and Global Production Networks in the Animation Industry." *Industrial and Corporate Change* 19, no. 1 (February 2010): 239–71. https://doi.org/10.1093/icc/dtp040.

Zook, Matthew. "The Geographies of the Internet." *Annual Review of Information Science and Technology* 40 (2006): 53–78. https://doi.org/10.1002/aris.1440400109.

———. *The Geography of the Internet Industry: Venture Capital, Dot-Coms, and Local Knowledge*. Malden, MA: Blackwell, 2005.

Zook, Matthew, Martin Dodge, Yuko Aoyama, and Anthony Townsend. "New Digital Geographies: Information, Communication and Place," edited by S. D. Brunn, S. L. Cutter, and J. W. Harrington, 155–76. Dordrecht, The Netherlands: Kluwer Academic Publications, 2004.

Zook, Matthew, and Mark Graham. "From Cyberspace to DigiPlace: Visibility in an Age of Information and Mobility." In *Societies and Cities in the Age of Instant Access*, edited by Harvey J. Miller, 241–54. Dordrecht: Springer, 2007.

Zuboff, Shoshana. *The Age of Surveillance Capitalism: The Fight for a Human Future at the New Frontier of Power*. New York: PublicAffairs, 2020.

Index

Abramitzky, Ran, 62
Advanced Research Projects Agency
    (ARPA), 71
advertising: as central pillar of digital
    capitalism, 2, 12, 21, 23, 25, 38–39,
    45–46, 54, 66; on Google Maps,
    development and evolution of, 50–53,
    51*fig*; Google PageRank and, 49, 50; IP
    (Internet Protocol) addresses and, 76; on
    MapQuest *vs.* Google Maps, 34, 35,
    45–48, 52, 53; targeted advertising, 25,
    29, 50, 66, 76, 84, 139
Aerojet Rocketdyne, 97–98
agriculture: as fundamental economic
    activity, 3–4, 23; Landsat and, 105;
    PlanetScope constellation, 86. *See also*
    food preparation and delivery
Airbnb, 2, 27, 77
Alaska Satellite Facility, 91
Algeria, 98
Alliance of Motion Picture and Television
    Producers (AMPTP), 3
allocation of goods and services: identity-
    location-transaction link, 77–84, 81*fig*,
    82*fig*; location and, 57
Alphabet, 46, 113. *See also* Google
Altavista, 49
Amazon, 1–2, 24, 54

Amazon Kindle, 2
American Automobile Association, 42
America Online, 43
angel investors, 11
antitrust regulations, 27, 135
Aoyama, Yuko, 21
Apple, 2, 48
Apple iTunes, 2
Apple Maps, 33; automobility and, 120;
    maps on the web and logic of digital
    capitalism, 33, 34–40
Apple TV+, 2
application programming interface (API), 53
Arab Spring, 27
Armenia, 88
ARPA (Advanced Research Projects
    Agency), 71
ARPANET, 70–73, 71*fig*, 72*map*, 86,
    135–36, 156n28
Artemis missions, 96–98
artificial intelligence (AI), 138; 2023 writer,
    actor, and producer strike, 3; satellite
    data and, 86, 88–89
Ask Jeeves, 49
Australia, 102–3*fig*, 104
Automatic Identification System, 87
automobility, system of, 115–21, 167n22;
    location, valuation, and marketization

Founded in 1893,
UNIVERSITY OF CALIFORNIA PRESS
publishes bold, progressive books and journals
on topics in the arts, humanities, social sciences,
and natural sciences—with a focus on social
justice issues—that inspire thought and action
among readers worldwide.

The UC PRESS FOUNDATION
raises funds to uphold the press's vital role
as an independent, nonprofit publisher, and
receives philanthropic support from a wide
range of individuals and institutions—and from
committed readers like you. To learn more, visit
ucpress.edu/supportus.

Milton Keynes UK
Ingram Content Group UK Ltd.
UKHW010844020724
444908UK00004B/71